Operation: INFINITE POTENTIAL

Restructuring the Energy Portfolio

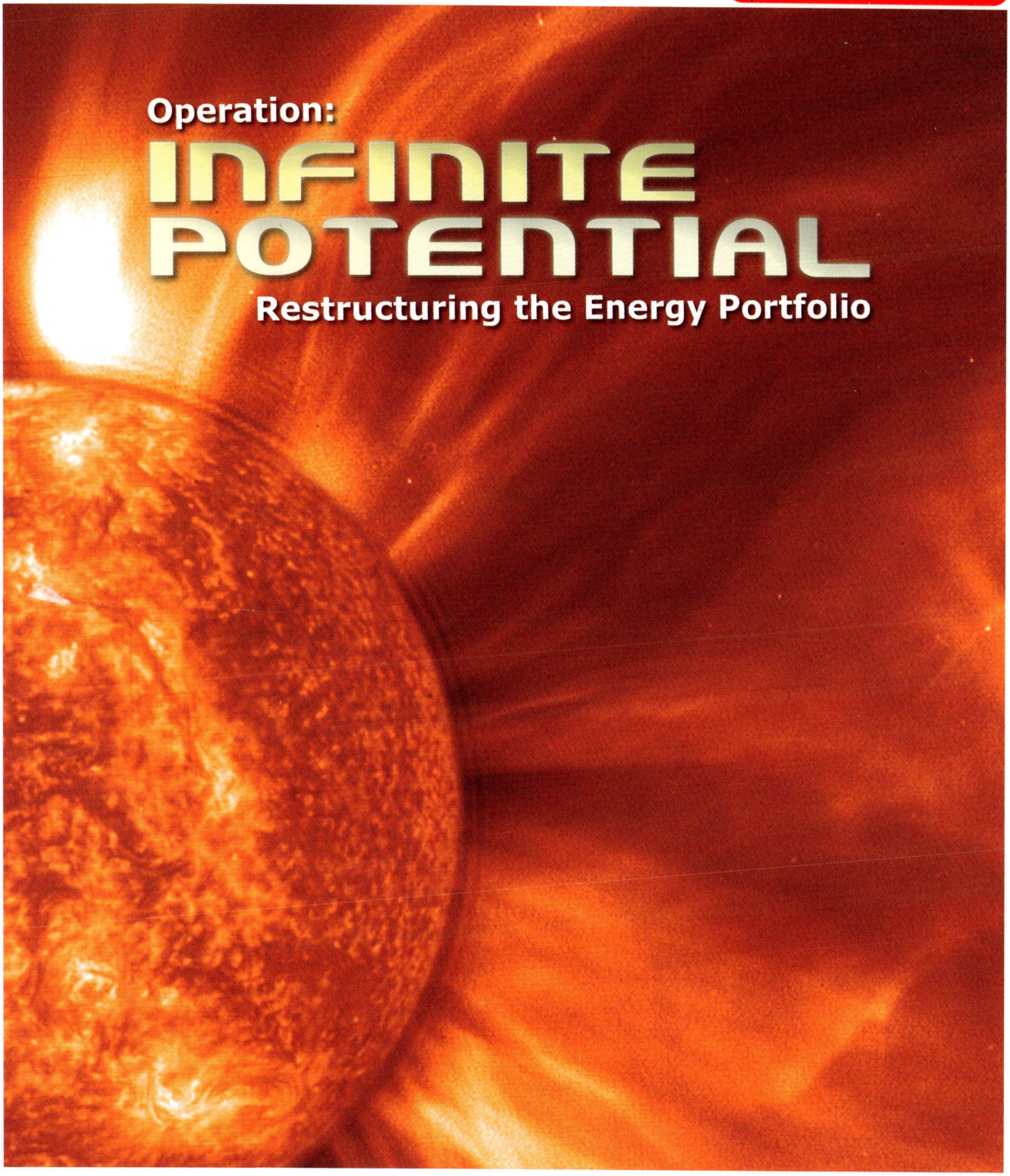

NATIONAL GEOGRAPHIC

The JASON Project℠

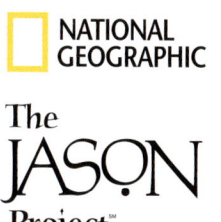

A nonprofit subsidiary of the National Geographic Society, The JASON Project connects students with great explorers and great events to inspire and motivate them to learn science. JASON works with the National Geographic Society, the National Oceanic and Atmospheric Administration (NOAA), the National Aeronautics and Space Administration (NASA), the U.S. Department of Energy, and other leading organizations to develop multimedia science curricula based on their cutting-edge missions of exploration and discovery. By providing educators with those same inspirational experiences—and giving them the tools and resources to improve science teaching—JASON seeks to reenergize them for a lasting, positive impact on students.

Visit *www.jason.org* to learn more about The JASON Project, or email us at *info@jason.org*.

Cover Design: Ryan Kincade, The JASON Project, and Navta Associates, Inc.

Cover Images

Main cover and title page: NASA.

Front cover thumbnails: (top) Student Argonauts get a true understanding of energy transfer and transformation on a roller coaster ride. Photo by Christine Arnold, The JASON Project; (bottom) Teacher Argonaut Bryan Ie and Student Argonaut Tim West are suited up and ready to enter the power plant. Photo by Peter Haydock, The JASON Project.

Back cover thumbnails: (top left) Student Argonaut Hannah Zierden. Photo by Peter Haydock, The JASON Project; (middle right) Student Argonaut Madhu Ramankutty on her way to the power plant. Photo by Peter Haydock, The JASON Project; (bottom left) The Argonauts. Photo by Peter Haydock, The JASON Project; (bottom right) Albert Einstein. Photo by Oren Jack Turner; Argonauts looking at the sky. Photo by Peter Haydock, The JASON Project; Water drop in a pond. Photo by Roger McLassus/Wikimedia Commons.

Published by The JASON Project.

© **2009 The JASON Project.** All rights reserved. No part of this publication may be kept in any information storage or retrieval system, transmitted, or reproduced in any form or by any means, electronic or mechanical, without the prior written consent of The JASON Project.

Requests for permission to copy or distribute any part of this work should be addressed to

The JASON Project
Permissions Requests
44983 Knoll Square
Ashburn, VA 20147

Phone: 888-527-6600
Fax: 877-370-8988

ISBN 978-0-9787574-9-6

Printed in the United States of America
by the Courier Companies, Inc.

10 9 8 7 6 5 4 3 2 1

National Geographic and the Yellow Border are trademarks of the National Geographic Society.

Contents

2 Getting Started with *Operation: Infinite Potential*

4 Your Tour of the *JASON* Mission Center

Missions

6 **Mission 1:** *Critical Current—Defining Energy*
Explore energy in its many forms. Lab 1 Energy Survey Lab **11** • Lab 2 Changes in Potential **15** • Lab 3 Exploring Visible Light **17** • Lab 4 Detecting Ultraviolet Radiation **19** • Field Assignment: Exploring Energy **24**

28 **Mission 2:** *Waves of Change—Calculating Transfers and Transformations*
Understand how energy is transferred and transformed in predictable ways. Lab 1 Energy Transfers and Transformations **32** • Lab 2 Wave Tank Tsunami **37** • Lab 3 Thermal Energy Survey Lab **39** • Lab 4 Electrochemical Cells **41** • Field Assignment: Exploring Energy Transfers and Transformations **44**

48 **Mission 3:** *Power to the People—The Current State of the Grid*
Investigate how we currently meet our energy and power needs. Lab 1 Exploring Magnetism **55** • Lab 2 Series and Parallel Circuits **57** • Lab 3 Generating Electricity **59** • Lab 4 Water Wheel **63** • Field Assignment: Don't Leave Footprints **66**

70 **Mission 4:** *Energy Independence—The Quest for Sustainable Resources*
Evaluate the future role of alternative energy resources. Lab 1 Wind Power **75** • Lab 2 Generating Hydrogen Gas **77** • Lab 3 Biofuels: Into the Woods **81** • Lab 4 Biofuels: Into the Lab **84** • Field Assignment: Enzymes Are Key **86**

90 **Mission 5:** *Energy Security—Powering Our Future*
Create a blueprint for a secure energy future. Lab 1 Cooling Off **95** • Lab 2 Making Models **99** • Lab 3 Using the Sun's Power at Night **101** • Lab 4 Communicating with Graphics **104** • Field Assignment: Commencing Countdown **106**

Connections

26 **History**
Sun Worship

46 **Math in Sports**
Coefficient of Restitution

68 **Culture**
Appalachia

88 **Renewable Resources**
Harnessing the Power of Plants

108 **Weird & Wacky Science**
Scientists Take Aim at Creating a Pea-Sized Sun!

Features

110 The JASON Project Argonaut Program

112 Meet the Team

114 Math and Building Tools

116 Glossary

Getting Started with *Operation: Infinite Potential*

Developed in collaboration with our partners at National Geographic, NOAA, NASA, the U.S. Department of Energy, and other leading organizations, *Operation: Infinite Potential* is built on a Mission framework to capture the energy and excitement of authentic exploration and discovery. The *Operation* consists of five captivating Missions that provide the real-world challenges, the science background knowledge, and the tools to help you solve each Mission challenge.

Let's take a closer look at the parts of each Mission!

Mission Objectives
Each Mission starts with a list of objectives that you will find on this opening page.

Mystery Connection
Each Mission presents a Mystery Connection. This is a problem to solve with clues you find while participating in the Mission.

Join The Team
Your Mission begins with an invitation to the join the Host Researcher and Argonaut team. You will work side-by-side with this team as they guide you through your study of energy.

Video and Online Resources
You will also see icons directing you to the Host Researcher Video where you will get to know more about the Mission team leader. Keep watch for these icons and others. They indicate when you will find multimedia resources online in the **JASON Mission Center.**

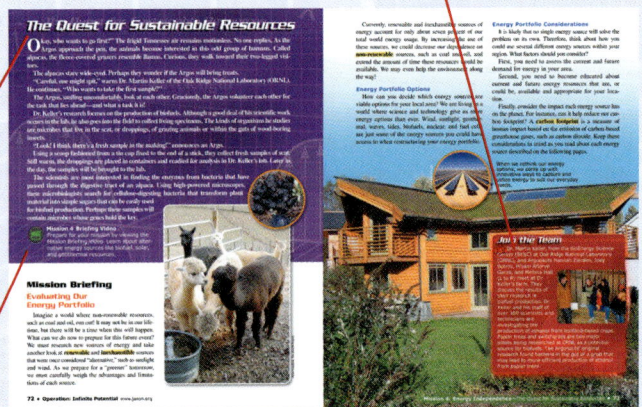

Introduction Article
Once you have your objectives and have met the team, each new Mission will introduce you to a day in the life of the Host Researcher, and the unique work that brings this scientist face-to-face with energy concepts.

Mission Briefing Video
See these adventures come alive in every Mission Briefing Video, which gives an action-packed introduction to the Mission objectives and key science concepts.

Briefing Articles
Gather all of your background information and clues through a series of Mission Briefing articles that guide you through the science of energy so that you can complete your Mission objectives.

Full-color graphics enhance the description and explanation of essential science concepts so you can clearly see the ideas presented in the briefings.

Researcher Tools
Check out the amazing tools that researchers use during their explorations in the field. From supercomputers to Laser Doppler Velocimeters, you will learn how these tools help researchers unlock the secrets of energy.

Fast Facts and Examples
You will find interesting things you have never thought of before in Fast Facts and Examples.

2 • Operation: Infinite Potential   www.jason.org

Team Highlights
Get an up-close view of the investigations that our Host Researchers and Argonauts have conducted during their field work for *Operation: Infinite Potential*.

International Connection
Check out examples of what is happening around the world and how scientists are exploring energy concepts.

Mission Labs
Put your knowledge to work with several hands-on labs in each Mission. The labs provide opportunities to practice and refine the skills you need in order to complete your mission objectives. In these labs, you will build tools, conduct investigations, collect data, and describe your observations and conclusions in your JASON Journal.

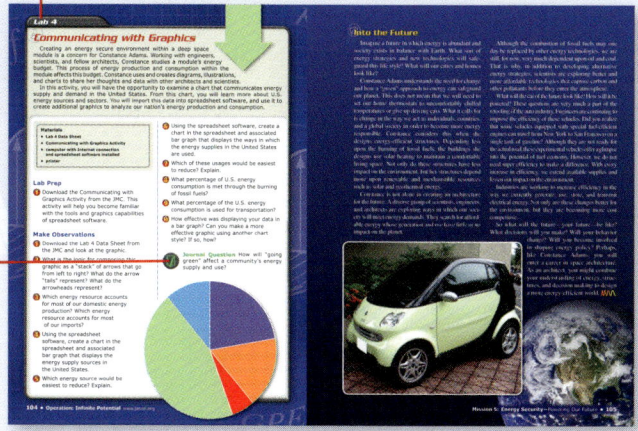

Additional Online Resources
You will find other great resources in the **JASON Mission Center** too, including your JASON Journal. Use it to record your work and experiences as you complete your *Operation: Infinite Potential* Missions. Check the JASON site often for live events and Webcasts that will provide updates on your Mission and on other breaking news in science.

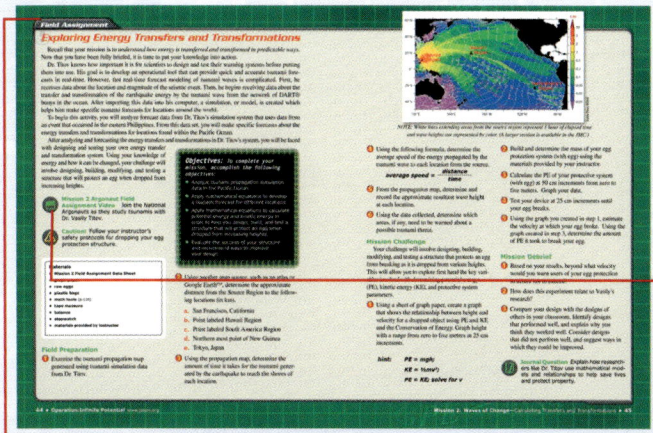

Connections
Learn to look for and find the amazing connections between science concepts and other things that you experience in the world around you. *Connections* highlight thought-provoking links that you can explore between science and human culture, history, geography, math, literature, strange phenomena, and other interesting topics.

Field Assignment
Field Assignments at the conclusion of each Mission give you the opportunity to put your new science skills and ideas to work in the field. To complete your Mission, you will need to accomplish the objectives set out in a Mission Challenge, and then provide an analysis during your Mission Debrief.

Argonaut Videos, Journals, and Photo Galleries
Argonaut Field Assignment Videos let you join the Argo team as they conduct their field work for selected Missions around the country. Login to the **JASON Mission Center** to read the Argonaut journals and take a look at their photo galleries documenting their field experiences!

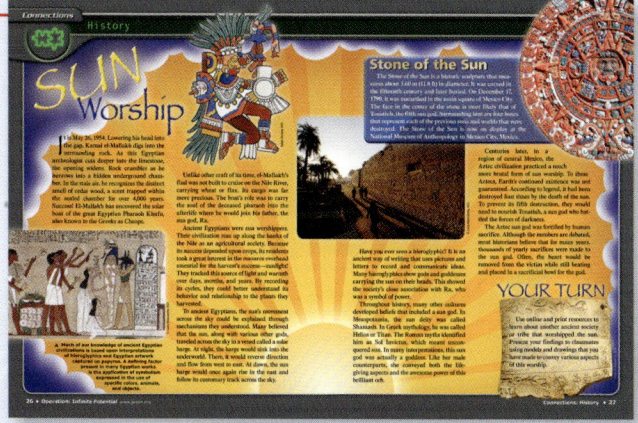

Getting Started with *Operation: Infinite Potential* • 3

Your Tour of the JASON Mission Center

The **JASON Mission Center** is your online hub for *Operation: Infinite Potential* content and resources and for the Argonaut community. Your JASON experience will come to life through interactive games, digital labs, video segments, your own JASON Journals, and other community resources and tools that support the Missions in this book.

Create Your Own Free Student Account

If your teacher has made an account for you, simply login to the **JASON Mission Center.** Otherwise, follow these simple steps below to create your own account.

1. Go to *www.jason.org*.
2. Look for the **JASON Mission Center** in the upper right corner.
3. Click "Register."
4. Choose "Student" as your role –OR– if your teacher provided you with a classroom code, enter it now.
5. Enter your email address or your guardian's email address and select a password for your account that you can easily remember.

The JASON Mission Center Home Page

Welcome to your **JASON Mission Center** home page. From here you can quickly access all the wonderful JASON tools and resources as you begin your mission. Take a moment to read the latest JASON news, try a search of the Digital Library, or jump right into *Operation: Infinite Potential* on the Web!

Here are some of the things you will see . . .

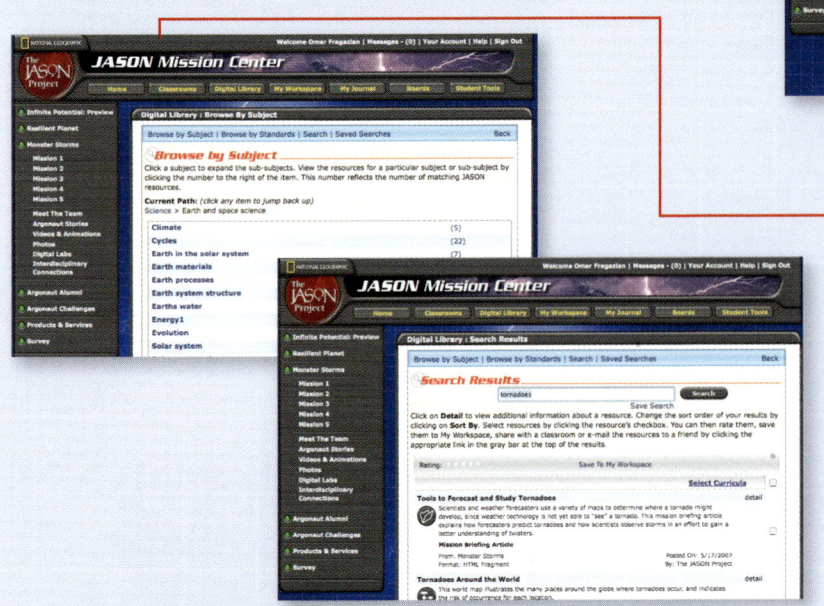

Your Resources and Tools

Powerful online tools are always at your fingertips. Use the *Digital Library* to find any JASON resource quickly and easily. Save and organize your favorites in *My Workspace*. View assignments and community updates in your *Classrooms* menu. These resources and more are always accessible through the *Tools* menu at the top of the **JASON Mission Center** page.

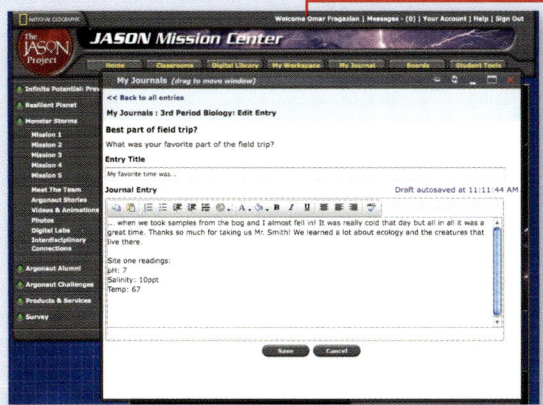

My Journals and Other Community Tools

Your student account on the **JASON Mission Center** includes an online JASON Journal that allows you to take notes, write about what you have learned, and respond to journal questions during the Missions. Other Community Tools include a moderated message board, classroom home pages, and tools to communicate with JASON researchers about their ongoing work in the field.

Online Version of Operation: Infinite Potential

This entire student edition book is also available to you online, for easy access anytime, anywhere. You can view any page from any Mission.

Team Info, Videos, and Photo Galleries

Learn more about the Host Researchers and the Argonauts from their biographies and journals. Video segments feature the Mission teams in action. Photo galleries provide additional views of the researchers and Argonauts at work, as well as stunning collections of more energy concepts at work.

Interactive Games

Visit the JASON Mission Center for digital labs, explorations, and games. See if you can build the most exciting roller coaster using your knowledge of potential and kinetic energy, and learn how tsunamis form!

Argonaut Resources

The Argonaut Challenge provides you and every other local Argonaut the chance to compete in a multimedia project that you can share online with the entire JASON community. You can also visit the message boards to discuss JASON with other local and National Argonauts around the world.

Your Mission begins at www.jason.org

Your Tour of the JASON Mission Center • 5

Mission 1: Critical Current
Defining Energy

"During a solar maximum, which occurs about every 11 years, the solar magnetic field flips and becomes twisted and very chaotic. During this time, we see sunspots, active regions, and coronal mass ejections that Earth has to contend with."

—Dr. Janet Green
Space Weather Physicist, NOAA

Janet Green

Janet Green, from the National Weather Service's Space Weather Prediction Center, uses data from satellites to protect Earth's array of satellites, astronauts, power grids, and airplanes from high-energy particles emitted by the sun.

 Meet the Researchers Video
Discover how scientists like Janet study the energy output from the sun and forecast the weather in space.

Space Weather Physicist, NOAA

 Read more about Janet online in the JASON Mission Center.

Photos above (left to right) NASA; James P. Blair, NGS; Michael Maggs/Wikimedia Commons; Peter Haydock, The JASON Project; Ken Bosman/Wikimedia Commons; NASA; Mike Lehmann/Wikimedia Commons; NASA; Alain Carpentier/Wikimedia Commons

Your Mission...

Explore energy in its many forms.

To accomplish your mission successfully, you will need to

- Discover the forms of potential and kinetic energy.
- Explain the benefits and risks of the different energy forms.
- Understand the characteristics and behavior of electromagnetic radiation.
- Evaluate the sun as a major source of energy in the solar system.
- Survey the particles and electromagnetic waves emitted by the sun.
- Examine how light interacts with matter.

Mystery Connection

ALERT!
To: JASON Space Weather Prediction Center

Space Weather
Current Conditions
Solar wind:
   Speed: 400.0 km/s (248.5 mi/s)
   Density: 7.80 protons/cm^3 (128 protons/in.^{3})

Space Weather Bulletin! An explosion within the sun's corona was detected earlier today. The major solar wind stream is following a collision course with Earth and is likely to produce a severe geomagnetic storm in three to four days.

Based upon this information, what should you do? Who should be alerted and what plans of action would you suggest?

Mission 1: Critical Current—Defining Energy • 7

Defining Energy

All eyes are focused on the displayed image of the sun. Its unnatural green color adds to the unreal setting. No doubt about it, this control center resembles a scene in a futuristic, sci-fi movie. The overhead monitors continually update images and graphics. Around the room, space weather specialists and other scientists keep careful watch on the data. Above the clamor of ringing phones, the scientists discuss the actions they need to take.

Things change. Minutes earlier, all was quiet. Space weather forecasters sat at their stations, routinely monitoring an unchanging stream of data. Then, it happened. Without warning, a region of the sun's surface violently exploded in a solar flare. In orbit above Earth, the Geostationary Operational Environmental Satellite (GOES) detected the event as a sudden increase in x-ray radiation. Data collected by satellites streamed down to this secure location in the foothills of the Rocky Mountains.

In addition to monitoring x-rays, the space weather specialists are also analyzing the flare that is most often followed by a **coronal mass ejection** (CME). A CME is a magnetic bubble or twisted rope of **magnetic field** lines that lift off from the sun. If this solar ejection reaches Earth's magnetic field, there could be severe and costly damage across the globe.

There are also other eyes on the data. Here, at the National Oceanic and Atmospheric Administration (NOAA) Space Weather Prediction Center (SWPC), Dr. Janet Green, a space weather physicist, observes these same readings on her desktop computer. Along with the other scientists, she is interested in how the sun's energy affects our planet.

 Mission 1 Briefing Video Prepare for your mission by viewing this briefing on your objectives. Learn how scientists like Dr. Janet Green need to understand energy in all its forms to predict the weather in space and its impact on Earth.

Mission Briefing

Energy

We have all heard the term *energy,* but what does it really mean? **Energy** is defined as "the ability to do **work**." In other words, it is the ability to move an object using a **force**.

Energy can be organized into two categories: potential energy and kinetic energy. As the name suggests, **potential energy** (PE) is stored energy. It is energy that is not yet in motion. It does, however, have the ability to be transformed into kinetic energy. By contrast, **kinetic energy** (KE) is the energy of motion.

Example

A snowboarder or cyclist stopped at the top of a hill has **potential energy.** Although she is not yet in motion, she has the potential to move. Once she shifts her balance and begins moving, she gains **kinetic energy** as she races down the hill.

8 • Operation: Infinite Potential www.jason.org

Forms of Potential Energy

There are many different forms of potential energy all around us. The book on your desk and the food you ate for breakfast represent just a few forms of potential energy. All forms of potential energy have the stored ability to move matter.

Gravitational

Gravity is a force of attraction that exists between any two objects. Gravitational potential energy is stored energy that depends on three things: the mass of an object, the height or potential falling distance of that object, and the acceleration of the object due to gravity.

An apple hanging on a tree has gravitational potential energy because it is being pulled toward a very large object — Earth. It has the potential to fall a distance. The higher up on the tree and the greater its mass, the more gravitational potential energy it has.

Gravitational Potential Energy

$$PE = mgh$$

PE = potential energy (in joules)
m = mass (in kilograms)
g = acceleration due to gravity (in meters/seconds²)
h = height (in meters)

Elastic

Stretch a rubber band. Now release it. What happens? The rubber band snaps back to its original length. The energy that powered this rapid return to the shorter, original length is another form of stored energy called elastic potential energy. As you stretched the band, you changed its shape. When the ends of the band were released, this stored energy was transformed into the "snap," or movement back into the original, more stable shape.

Example

Elastic potential energy is also found in another common action — bouncing a ball. When a round ball strikes a hard surface, the round ball flattens where it hits the surface. Even though the change may be too slight and quick to observe, striking the surface alters the tension in the ball's material. Like a stretched rubber band snapping back to form, the ball returns to its round appearance. As it reshapes, it pushes off the hard surface, resulting in a bounce.

In the 17th century, Sir Isaac Newton described gravity as a force of attraction and devised equations to explain its behavior. Over 200 years later, Albert Einstein theorized that gravity warped space and produced distorted paths along which objects moved as illustrated at right. Today gravity remains a mystery. An alternate theory proposes the existence of tiny, gravity-producing particles called gravitons. Perhaps in the future, you will be the scientist who will solve the mystery of gravity.

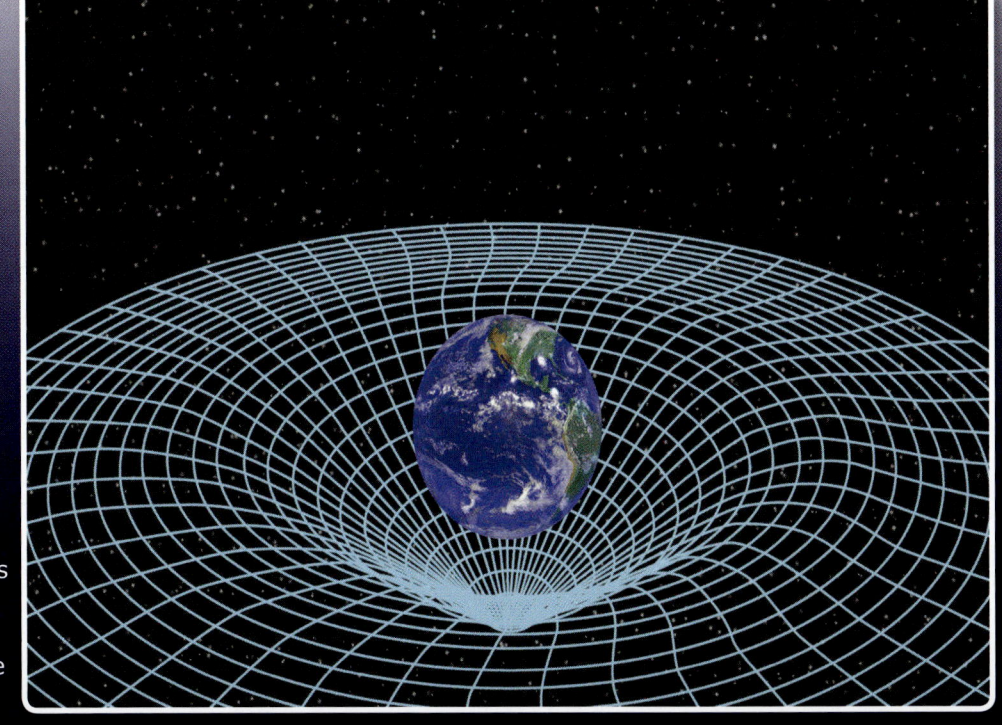

Mission 1: Critical Current—Defining Energy

Magnetic

Bring two magnets together. Depending on which ends face each other, the magnets will either pull together or push apart. This attraction or repulsion is a result of the magnetic potential energy stored in these objects. This same magnetic potential energy is found in the coronal mass ejections (CMEs) studied by Dr. Janet Green. CMEs carry with them a part of the sun's magnetic field, impacting our planet in many ways.

Electrostatic

Another form of potential energy, electrostatic potential energy, arises from the attraction and repulsion of electrical charges and occurs when certain materials are brought together. If you have ever had a clean sock stick to your clothes after it was removed from the dryer, you have experienced the effects of electrostatic potential energy.

Chemical

Chemical potential energy is energy stored within the bonds of a substance. Although you cannot see the stored energy, you can experience its effect when substances react.

In any chemical reaction, **Conservation of Mass** is maintained. Although bonds are broken and new substances are formed, the total amount of matter does not change.

▲ All chemical bonds contain chemical potential energy. Consider the food we eat. The bonds that form these food molecules contain potential energy. Living cells use chemical reactions to break these bonds. As the food is broken down, energy from the bonds is released. This powers the chemical processes that keep our cells alive.

Example

Chemical potential energy exists within the chemical bonds of substances, such as gasoline. Ignited by a spark in the car's engine, gasoline molecules react with oxygen, producing an explosive reaction. During this process, chemical energy that had been stored in the bonds of the fuel molecules is released. Through a series of energy transfers and transformations, the vehicle moves. The Conservation of Mass is maintained because the mass of the gasoline and oxygen that react equals the mass of carbon dioxide and other emissions produced.

Nuclear

Nuclear potential energy is the energy stored in the bonds of subatomic particles. The atom's nucleus contains most of this energy. This energy can be released by either splitting or fusing atoms through the processes of fission and fusion. In these reactions, some mass is transformed into energy.

Join the Team

In Boulder, Colorado, at the NOAA Space Weather Prediction Center, the Argonauts and Dr. Janet Green observe the satellite receivers that collect space weather data. Dr. Green analyzes the data to determine when coronal mass ejections might have an effect on satellite systems here on Earth. Dr. Janet Green, Tom DeFoor, Lindsay Hannah, Joey Botros, Bryan Ie, and Toba Faseru (L to R) learn that CMEs can affect satellites orbiting Earth, electronics and power grids, and radio communication systems, causing massive disruptions important to our everyday lives.

10 • Operation: Infinite Potential   www.jason.org

Lab 1

Energy Survey Lab

When researching and forecasting solar events that could impact Earth, Janet Green must observe and analyze a variety of energy forms, including the electromagnetic energy in x-rays, the magnetic energy of coronal mass ejections, and the kinetic energy of protons. Janet must monitor these energy forms to make reliable forecasts that she can communicate to airlines and power companies. Even though you may not be familiar with the science of energy, you have experienced it in many forms. Flip on a switch and a lamp lights up. Talk aloud and your voice carries across the room. Pedal your bike and energy moves you forward.

In this activity, you will be challenged to take a closer look at energy forms. Through observation, inquiry, and analysis, you will develop a better understanding of energy. This will be accomplished as you move from station to station and critically explore setups that present an assortment of energy concepts.

Materials
- Lab 1 Data Sheet
- materials provided by instructor

Lab Prep

1. When you enter the laboratory, do not touch any of the stations. Wait for instructions from your instructor before beginning the lab.
2. With your instructor, review all appropriate laboratory procedures, safety guidelines, and classroom rules.
3. Review the objectives and any procedures that are established for each station.
4. Review the order in which you will move from station to station.

Make Observations

1. After your instructor has presented the lab prep, a signal will be given to begin work at your first station. Remember that you only have a limited time to work at each station and answer questions.
2. Use whatever tools and materials that are available at that station to perform the suggested investigations.

Journal Question Try to list all of the forms of energy that you encounter during a typical day. Explain how you interact with each energy form.

Team Highlight

Dr. Janet Green, Joey Botros, Toba Faseru, Lindsay Hannah and Teacher Argonaut Bryan Ie (L to R) test for the presence of ultraviolet (UV) radiation. UV radiation can have damaging effects to our skin and eyes. The Argos investigate the properties of UV radiation in order to learn how to protect our bodies.

Mission 1: Critical Current—Defining Energy • 11

Forms of Kinetic Energy

There are many forms of kinetic energy (KE) around us. We can easily observe some forms of kinetic energy in moving vehicles, passing clouds, and moving planets and stars. However, some forms of kinetic energy cannot be observed directly. For example, the flow of electrical energy in a wire is not directly observed, but we can see its effect when we turn on a light.

Mechanical

The energy of an object in motion is called mechanical energy. Mathematical formulas can help calculate the mechanical energy found in moving objects. The amount of mechanical energy depends upon two things—mass and velocity. The larger the mass or greater the velocity, the more mechanical kinetic energy possessed by the object in motion.

> **Mechanical Kinetic Energy**
>
> $$KE = \tfrac{1}{2}mv^2$$
>
> **KE = kinetic energy** (in joules)
> **m = mass** (in kilograms)
> **v = velocity** (in meters/seconds)

Dr. Green studies protons associated with coronal mass ejections. Even though these tiny particles have a very low mass, they move at incredibly high velocities, giving them a lot of kinetic energy.

In space, the impact by just one of these high-energy particles can have serious effects on spacecraft and their crews. These particles can destroy sensitive satellites or corrupt computer programs. They can even injure living cells and damage DNA.

Thermal

The particles that make up a substance are always vibrating, stretching, and rotating. **Thermal energy** is the total energy within a substance. As thermal energy in a substance is increased, the particles vibrate, stretch, and rotate faster, giving them the energy to move apart. If enough thermal energy is added to certain solid substances, they will eventually melt and become a liquid. Adding more thermal energy will eventually turn this liquid into a gas. When we measure a substance's **temperature**, we are measuring the concentration of thermal energy.

Electrical

Electrical energy is kinetic energy that results from the movement of charges. There are electrical wires running through your home. When you flip a light switch on, you allow electrical energy to flow through the wires to the light bulb.

Sound

When you hear music coming out of your radio, you feel vibrations that are associated with another form of kinetic energy—sound energy. The speaker in your radio vibrates the air. These vibrations pass along through the air causing your eardrums to vibrate. Our brain interprets these vibrations as sound.

Electromagnetic

Electromagnetic energy travels as **waves** and does not require a medium through which to travel. This form of energy can travel through the vacuum of space, and has electrical and magnetic properties. In a vacuum, **electromagnetic waves** move at the speed of light. Sunlight is an example of electromagnetic energy.

The sun is the primary source of most energy in our solar system. It heats Earth, provides energy for growing plants, and provides most of the visible light we use to see. Some of the sun's energy from thousands and millions of years ago was captured and stored as substances we use today for energy. Because the sun directly or indirectly provides so much energy, we will next look at its energy output and how it impacts our lives.

> **Duality of Light**
>
> Although light behaves as a wave, it also has properties of a particle. Scientists call basic units of light *photons*. The amount of energy contained by a photon is determined by its frequency. High frequency waves have more energetic photons than lower frequency waves.

Forms of Energy

Potential Energy

Gravitational

Energy that an object has due to its position in a gravitational field. Examples include an apple on a tree, a cyclist at the top of a hill, and a book on your desk.

Elastic

Energy stored in the bending, stretching, or twisting of an object. Examples include a bent bow, wound spring, or stretched elastic.

Chemical

Energy stored within the chemical bonds of a substance. Examples include the food we eat, gasoline, and a candle.

Nuclear

Energy stored in subatomic matter. An example includes the uranium rods found in power plants.

Magnetic
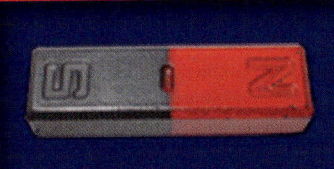
Energy stored within magnetic fields that can be seen in the attraction or repulsion of objects. Examples include magnets on your refrigerator and the needle on a compass.

Electrostatic

Energy that an object has due to its position in an electric field. Examples include clothes sticking together when they come out of the dryer or your hair sticking up after taking off a wool hat.

Kinetic Energy

Mechanical

Energy of motion that is attributed to a specific object. Examples include waterfalls, a car moving down the street, and a ball flying through the air.

Thermal

Total energy within a substance, measured in units of heat and temperature. Examples include ice melting and water boiling.

Electrical
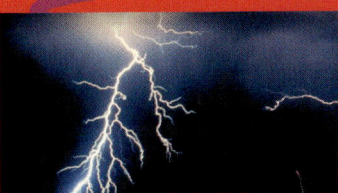
Energy associated with the movement of charges. Examples include lightning and a lit light bulb.

Sound

Energy that is transmitted through the compression of matter. Examples include hearing music from a radio and hearing thunder.

Electromagnetic
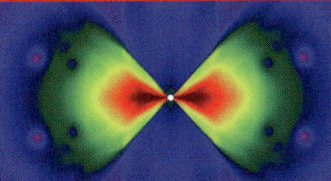
Energy that travels in waves and can travel in a vacuum. Examples include visible light, a rainbow, and radio waves.

Mission 1: Critical Current—Defining Energy • 13

Energy, Work, and Power

How do we measure energy? Both potential and kinetic energy can be measured in joules. A **joule** is the basic unit for energy or work.

Measuring Work

What is **work**? Work is the energy needed to move an object a certain distance using a **force**. Lift a ball into the air, and work is performed by your body's muscles. Let the ball drop onto the ground, and once again work has been done—this time by gravity. However, if you were to hold that ball above your head, keeping it at the same position, no work would be performed. Although your muscles may get tired from holding something up, if the ball's position does not change, no work has been performed on the ball.

Work

$$w = Fd$$

w = **work** (in joules)
F = **Force** (in Newtons)
d = **distance** (in meters)

Fast Fact

The term horsepower (hp) was first used as a way of comparing the power of early steam engines to that of horses. James Watt, after whom the unit of power is named, based the measurement of horsepower on the estimated amount of coal that could be lifted by a horse. He came up with a value of 33,000 foot pounds per minute, which converts into 746 joules per second, or 746 watts.

Measuring Power

When considering work, the rate at which it is performed is also important. **Power** is the rate at which work is performed, and is measured in terms of watts (W). One watt equals one joule per second. When you lifted the ball above your head, the work, or energy needed to lift the ball, is the same whether you lifted it slowly or quickly. However, you used more power when you quickly lifted the ball.

Power

$$P = \frac{w}{t}$$

P = **power** (in watts)
w = **work** (in joules)
t = **time** (in seconds)

Example

The **power** output of lamps is measured in watts. Lamps rated at different wattages use different amounts of energy over a given unit of time. When you turn on a 100-watt incandescent lamp, it uses 100 joules of energy every second it is on. Compare this to a 60-watt incandescent lamp which uses 60 joules of energy in one second. The 100-watt lamp uses more energy than the 60-watt lamp and consequently appears brighter.

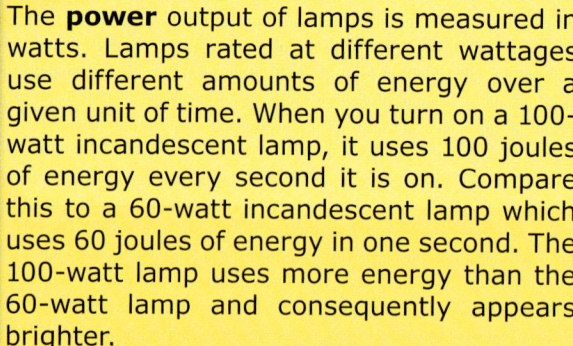

Gilbert M. Grosvenor, NGS

14 • Operation: Infinite Potential   www.jason.org

Lab 2

Changes in Potential

Have you ever wound up a rubber band toy, then released it? The stored energy of the twisted elastic transforms into the energy of the toy's motion. In a similar way, the sun's magnetic field stores energy in its twists and turns. Part of this twisted magnetic field can lift off in a burst called a coronal mass ejection, which Janet Green studies.

In this activity, you will analyze balloon rockets to determine the relationship between potential and kinetic energy transformations. As in a twisted magnetic field, a stretched balloon can store elastic potential energy when inflated. We can observe the effects of this energy as it is transformed into mechanical kinetic energy when the balloon is released. Using the formula $KE = \frac{1}{2} mv^2$ we can calculate its kinetic energy. From this value, we can infer the relative amount of elastic PE stored in the stretched rubber of the balloon.

Materials
- Lab 2 Data Sheet
- large balloons
- kite string
- scissors
- math tools (p.114)
- 2 chairs
- straws
- stopwatch
- meter stick
- safety goggles
- tape

 Review safety precautions with your instructor before beginning this lab. When working with projectiles, always wear goggles.

Lab Prep

1. Work in a large space such as a gym or multipurpose room.
2. Obtain a length of kite string about 10 m long. Pass the string through a drinking straw.
3. Tie each end of the string to chairs positioned at opposite sides of the room. Make sure the string is level and taut. Slide the straw to one end of the string.
4. Now inflate the balloon, but do not tie it. Use a strip of tape to attach the inflated balloon to the straw. Make sure that the nozzle points in the opposite direction of the intended path. Secure the balloon so that it hangs below the string.

Make Observations

1. Release the balloon and observe its behavior. Use the stopwatch to time how long the balloon rocket travels. Measure the distance traveled in meters. Record both values.
2. Using the math tools, calculate average speed. This value can be used to represent average velocity, as average velocity is the average speed with direction.
3. Obtain the mass of the balloon/straw system. Discuss the order of operations and use the math tools to determine the kinetic energy of the balloon.
4. If we assume no energy loss through transformation of elastic potential energy to kinetic energy, how much elastic potential energy was stored in the stretched rubber of the balloon? Explain.
5. Identify factors that influenced the potential energy and kinetic energy in your investigation. Explain using the data you collected.
6. Design an investigation that tests the influence of one of the factors you have identified.

Extension
Design an investigation that measures gravitational potential energy and kinetic energy of two or more objects that have been dropped.

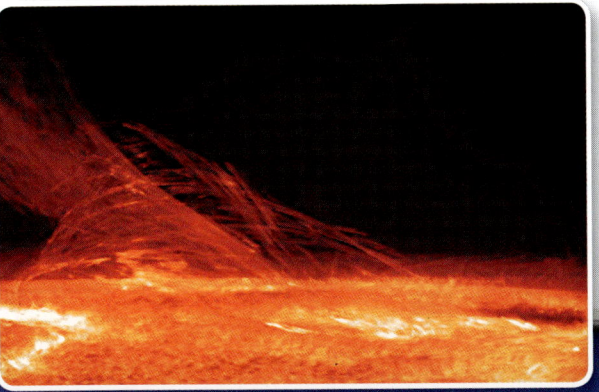

 Journal Question Imagine and draw a machine that could be powered by elastic potential energy.

Mission 1: Critical Current—Defining Energy

Electromagnetic Energy: Visible Light

Look around you. The images that you see are formed by visible light waves (electromagnetic energy) **reflecting** off the surface of objects. Your eyes detect these waves and communicate this information to your brain.

Light Rays

Electromagnetic energy travels in straight lines. This behavior of light is often illustrated in drawings called **ray diagrams.** In ray diagrams, a few representative waves are shown as lines with arrowheads, called rays.

Transmission, Reflection, Absorption

When light rays strike an object, several things can happen. The electromagnetic energy can be allowed to pass through—**transmitted**, or prevented from passing through—either **absorbed** or reflected. Sometimes, some of the light rays are transmitted, while others are absorbed and/or reflected.

Transparent, Translucent, Opaque

When materials allow light to pass through with little change to the pattern of rays, the material is considered **transparent.** Clear glass, some plastics like plastic wrap and sandwich bags, and some liquids like water and alcohol are transparent.

When the light is allowed to pass through, but the object distorts the pattern of the rays, the image on the other side may appear cloudy because some of the rays cannot reach our eyes in the original pattern. This material is called **translucent.** Inflated plastic balloons, some plastic storage containers, and rippled glass are all translucent.

When light is not absorbed and does not pass through an object, but reflects off its surface, the object is called **opaque.** Some of the light is absorbed and some reflects, and you can see the object, but nothing on the other side of the object. Your desk, a brick, and cardboard boxes are all opaque.

If light rays bounce off an object then strike another opaque object with a smooth surface, such as a mirror, the rays reflect back to our eyes in a parallel pattern. With the original ray pattern intact, we can see reflected images on the mirror.

Lenses

Lenses transmit light, but alter the direction of the electromagnetic energy. Some lenses concentrate light waves together, or focus light, while others disperse light. **Convex lenses** have a bulge in their center and **concave lenses** have a depression in their center. Lenses are used in microscopes, glasses, and cameras.

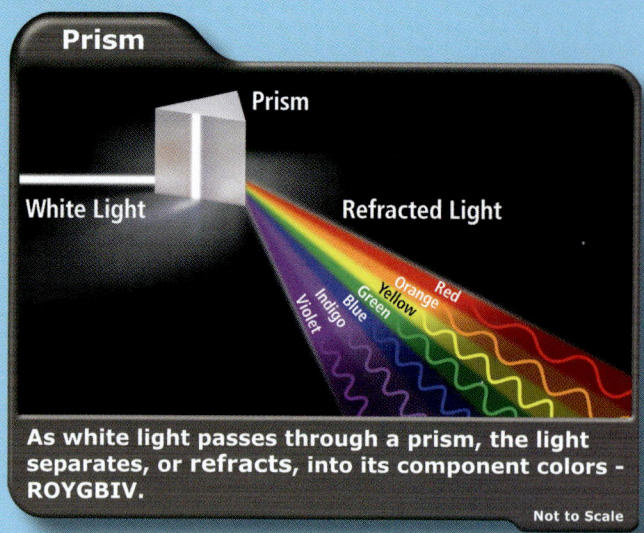

Prism

As white light passes through a prism, the light separates, or refracts, into its component colors - ROYGBIV.

Not to Scale

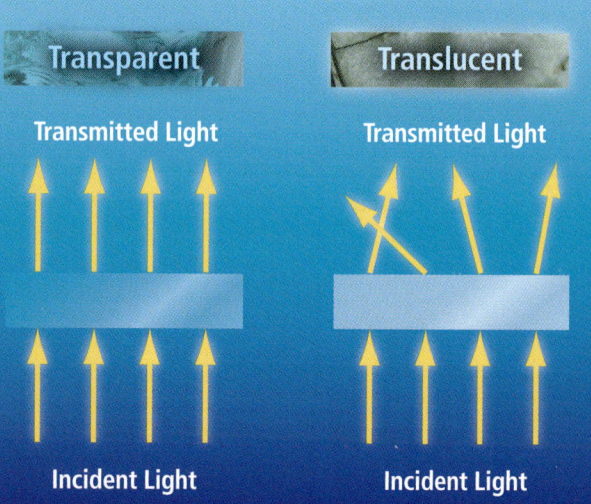

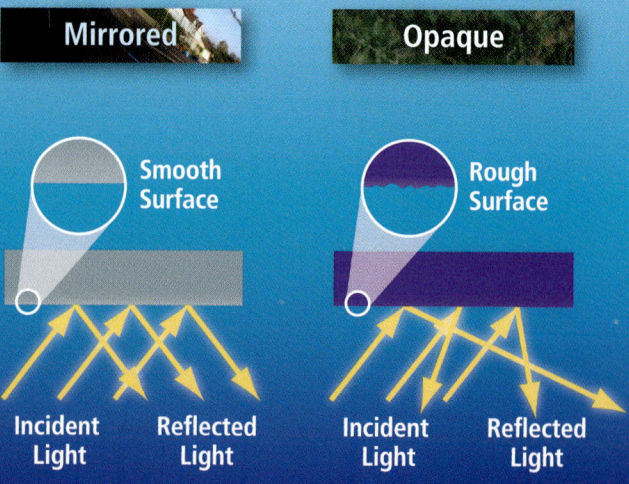

16 • Operation: Infinite Potential   www.jason.org

Lab 3

Exploring Visible Light

As the sun's electromagnetic energy travels through space, it interacts with different materials along its journey to and around Earth. Based on the properties of the material, electromagnetic energy will be affected in a variety of ways. Janet Green must observe, explore, and monitor these effects as she makes space weather forecasts.

In this activity, you will work through a set of investigations to observe and analyze how visible light behaves when it interacts with a variety of materials. Based on these interactions, you will analyze each experience and record what you observe in graphs and diagrams. The data will be used to help you illustrate and understand the behavior of light rays.

▲ The surface characteristics of an object determine the behavior of light rays that strike it.

Materials
- Lab 3 Data Sheet
- graph paper
- aluminum foil
- flashlight
- ruler
- comb
- modeling clay
- flat mirror
- flexible mirror
- concave lens
- convex lens

Lab Prep

1. Dim the lights in the room. Hold a flashlight 10 cm above a sheet of graph paper. Observe the characteristics of this beam, noting the circles of different brightness. Measure and record the diameter of the outermost beam.

2. Determine the relationship between the distance you hold your flashlight from the sheet of paper and the diameter of the circle of light that appears on the paper. Collect data and graph your results.

3. From your data and graph, explain the relationship between beam diameter and flashlight distance. Did all groups in your class discover the same association? Discover why or why not.

Make Observations

1. Separate two desks or lab tables, forming a gap of about 5 cm. Position a comb several centimeters away from and parallel to the gap. With its teeth pointing down, place a small lump of clay at either end of the comb to secure its upright position.

2. Shine the beam of a flashlight through the comb teeth, casting a shadow across the gap. What do you observe? How does the distance from light to comb affect the cast pattern? Explain.

3. Use aluminum foil to cover a good portion of the comb's teeth, producing a central, stamp-sized window that will limit the number of projected rays.

4. Adjust the flashlight's angle, position, and distance to produce a pattern of rays that appear parallel. Hold a mirror in this light-ray pattern. What happens to the rays that strike the mirror's surface? Draw a picture of what you observe.

5. Examine a convex and a concave lens. Make a sketch that illustrates a cross section, or side view, of each lens. Look through these lenses at nearby and distant objects. How do the lenses affect the viewed images? Does the distance to an object affect the transmitted image? Explain.

6. Place each lens within the table gap so that parallel rays strike the midsection of the lens. How does passing through each lens affect the rays? Make a ray diagram that illustrates your observations.

7. Obtain a small sheet of flexible mirror. Examine your reflection. How does flexing the mirror affect the reflected image of your face? Position the mirror within the comb's ray pattern and observe the angles at which rays are reflected. Does altering the mirror's reflective surface from concave to convex affect the observed rays? Explain. Make a ray diagram that illustrates your observations.

Journal Question Explain the role of different lenses that are found in your home, school, and neighborhood.

Mission 1: Critical Current—Defining Energy • 17

Electromagnetic Energy: Infrared and Ultraviolet

Visible light isn't the only type of electromagnetic wave. There's an entire spectrum of waves that come from the sun. They vary in the amount of energy contained in each.

Infrared

Infrared (IR) is electromagnetic energy with slightly less energy content than visible light. IR is invisible to most animals, including humans. Some animals, however, have evolved the ability to detect IR.

IR is also called thermal radiation. This is because exposure to this type of electromagnetic energy produces a notable rise in temperature.

> **Fast Fact**
>
> The ability to detect infrared radiation can sometimes be used by animals to "see" in the dark. This adaptation is found in certain snakes. However, what they "see" is not visible light. They detect the infrared radiation given off as thermal energy from warm-blooded animals. This gives the snake an advantage at night when hunting for prey, like rodents.

Ultraviolet

Invisible electromagnetic energy with slightly more energy content than visible light is called **ultraviolet** (UV). Like IR, UV radiation is invisible to humans. It can, however, be detected by some species of birds, reptiles, and insects. For example, bees can see UV patterns on flowers, leading them to pollen.

There are three main types of UV radiation. Distinguished as UVA, UVB, and UVC, these forms of ultraviolet radiation differ in energy content. UVA radiation contains the least energy and it comprises most of the UV radiation that passes from the sun through the atmosphere. This radiation can cause damage to connective tissue and increase a person's risk for skin cancer.

UVB contains sufficient energy to be quite harmful. Exposure to UVB can cause a sunburn. It can also produce changes in DNA, resulting in mutations

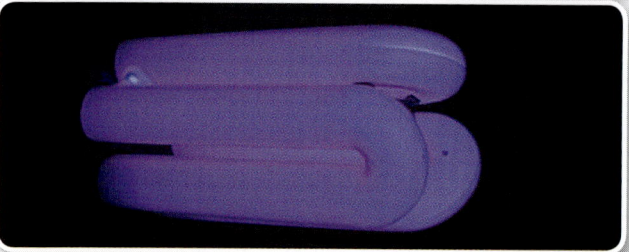
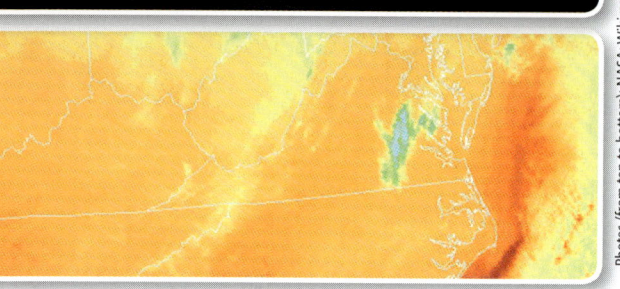

▲ Although our eyes cannot detect electromagnetic energy outside the visible spectrum, we can use this radiation in various ways. Heat or IR signatures of warm-bodied animals can be detected by cameras and displayed as eerie visuals. UV radiation, emitted by a black light, can energize special inks causing them to glow in the dark. Satellites using IR cameras can capture information about the planet's temperature, vegetation, and other ground characteristics.

Photos (from top to bottom): NASA; Wikimedia Commons; aviationweather.gov

in skin cells. Some of these mutations may lead to skin cancer.

UVC is the most energetic and damaging form of ultraviolet radiation. All of it, luckily, is screened out by the atmosphere before it reaches Earth's surface.

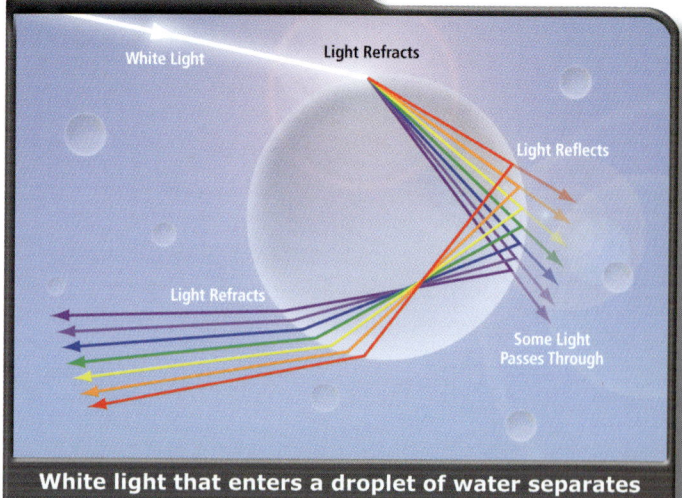

Rainbow Formation

White light that enters a droplet of water separates into its component colors and reflects out as a spread of wavelengths we observe as a rainbow.

Lab 4

Detecting Ultraviolet Radiation

Although Janet Green mostly studies the high-energy particles emitted by the sun, she is also aware of its electromagnetic radiation. Like high-energy particles, electromagnetic waves stream across the vacuum of space. Although an assortment of waves strike Earth's outer atmosphere, only narrow bands of this energy reach our planet's surface. Most electromagnetic radiation is absorbed by our atmosphere.

UV radiation contains slightly more energy than can be detected by the human eye. So although this energy strikes the eye's photosensitive layer, we do not detect it. This radiation, however, can be sensed by a variety of devices. In this activity, you will use inexpensive plastic indicators, known as UV beads, to detect ultraviolet radiation.

Materials
- Lab 4 Data Sheet
- UV beads
- shoe box
- ruler
- materials provided by instructor

Lab Prep

1. Review with your teacher what UV beads are and what they indicate.
2. Place some UV beads in one hand. Close this hand tightly, not allowing light to reach the beads.
3. Walk into a location bathed in direct sunlight.
4. Open your hand. Observe the beads. What happens? How long does it take? Did all the beads react in the same way? Write down your observations.
5. Shield the beads again from direct sunlight. Wait several minutes. Look at the beads now and describe any changes you observe.
6. Close your hand and return into the sunlight. Once again open your hand, exposing the beads to the direct rays of the sun. What did you observe?
7. With your instructor, discuss the behavior of these UV-detecting beads.
8. Turn a box on one side, so that the box opens horizontally.
9. Fasten UV beads in a line at the bottom far edge of a box, separating each from its neighbor by about 2.5 cm.

Make Observations

1. Transport the box outdoors into an area with direct sunlight, making sure that sunlight does not stream into the box. Position the box on a flat, horizontal surface so that the opening faces away from direct sunlight.
2. Design an investigation to determine if UV behaves like visible light, based on what you learned about visible light in *Lab 3: Exploring Visible Light*.
3. With your instructor's permission, perform your investigation and record your results.
4. Design an investigation in which you determine the effectiveness of various items at blocking UV light.
5. With your instructor's permission, perform your investigation and write down your results.
6. Once the investigations are completed, share your results with classmates. Based on what you have learned, compose additional questions that could be answered by your classmates.

 Journal Question How can you use these beads to help you protect yourself against UV radiation?

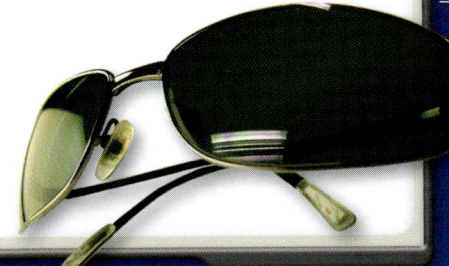

Mission 1: Critical Current—Defining Energy • 19

Electromagnetic Energy: The Ends of the Spectrum

Gamma and X-Rays

In addition to emitting energy in the forms of visible light, IR, and UV, the sun also emits higher energy gamma rays and x-rays and lower energy microwaves and radio waves. Gamma rays, x-rays, and a good deal of ultraviolet radiation are screened out in the Earth's stratosphere. As you move across the spectrum from gamma to radio waves, you encounter photons with decreasing energy content.

Example

Overexposure to **x-rays** is harmful because they can pass through many materials, including the tissue that makes up our bodies. However, medical technologies have allowed us to use the x-rays in limited exposure to look inside our bodies and treat diseases.

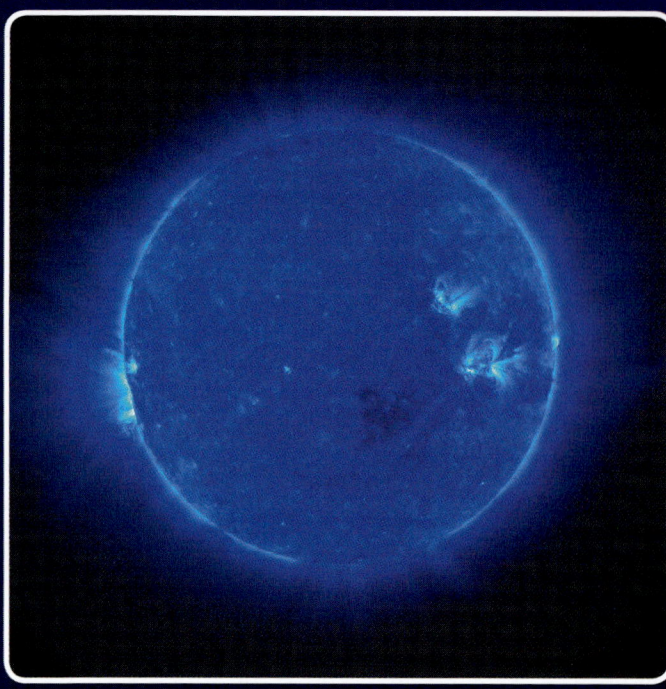

Telescopes and satellites collect data on the sun and stars using other parts of the spectrum, including microwaves, x-rays, and radio waves. This image is given a false color so that we can see areas that are emitting higher levels of these normally invisible forms of energy.

Microwaves and Radio Waves

Microwaves and radio waves have lower energy content than the rest of the electromagnetic spectrum. Like the other components of the spectrum, we have learned to make use of their unique properties in our every day lives.

You may be familiar with microwaves because many homes use this form of energy to cook or heat food. Radio waves have the lowest energy content of all the types of electromagnetic energy. This part of the electromagnetic spectrum is very useful in long distance communications. Radio stations use the properties of radio waves to broadcast signals spanning large distances at certain energy levels. Your radio can "tune" into these frequencies so you can listen to your favorite songs.

Here on Earth, astronomers can use radio telescopes to explore space because most radio waves are able to penetrate Earth's atmosphere. Using radio waves, astronomers are able to observe objects in deep space they could not see with a normal optical telescope.

Fast Fact

Scientists have launched many satellites in order to study the sun. One of these satellites, the Earth Radiation Budget Satellite (ERBS), was launched by NASA to measure the amount of energy transferred from the sun to Earth. In a balanced system, the amount of electromagnetic energy striking our planet equals the amount released back into space. These satellites are helping us to verify these amounts over a long-term study. These data are compiled into a radiation budget that conforms to the Conservation of Energy which we will learn more about in the next Mission.

The Radiation Budget

Changing the Budget Balance

These days, the radiation budget of the planet is changing. Nearly the same amount of electromagnetic radiation is being transferred to Earth, but the length of time it takes to leave Earth is increasing. This has produced a slow-but-steady rise in the average temperature of the oceans and atmosphere.

The Life Zone

We are in a region of space where the radiation budget creates a temperature range on the planet in which life as we know it can exist. Based on the energy output of the sun and the orbit of Earth, this region is called the life zone. Currently, Earth is the only planet in our solar system we think is in the life zone. However, in the past, other planets, such as Mars or Venus, may have also been in the life zone.

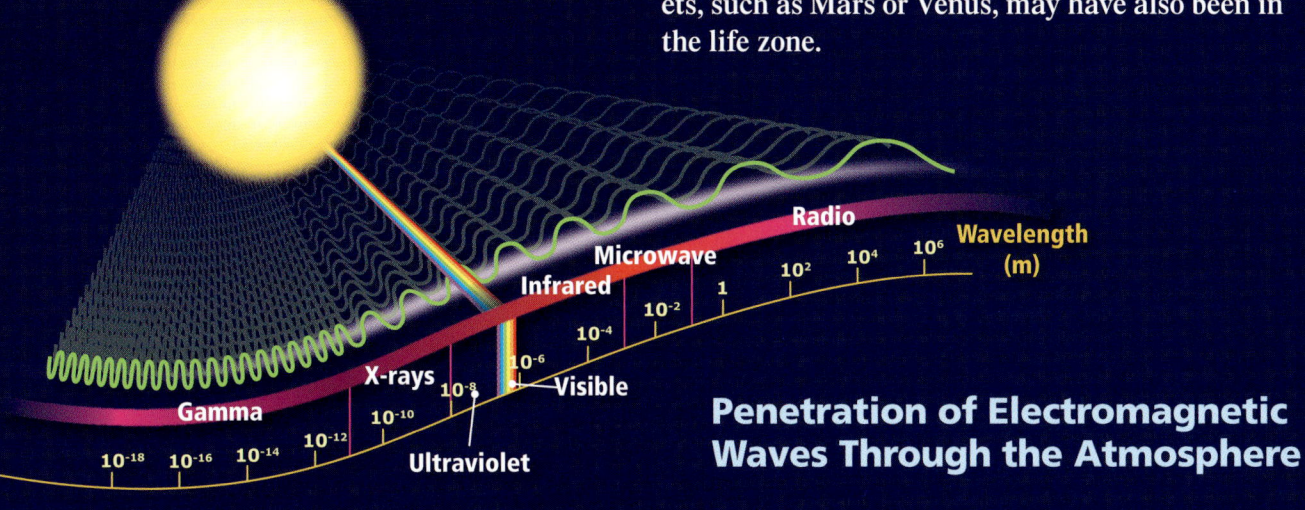

Ground-based telescopes are limited to detecting and analyzing only the parts of the electromagnetic spectrum that pass through the atmosphere. That is why surface observation is performed mostly by optical telescopes that explore the visible spectrum and radio telescopes that analyze radio waves. Janet Green uses satellites that are capable of detecting and measuring the sun's x-rays, which are an indicator that a coronal mass ejection and energetic photons are on their way. These are important to detect because they can damage satellites and shut down electrical grids.

Mission 1: Critical Current—Defining Energy • 21

Energy from the Sun

Solar Wind

The sun is the original source of most energy on Earth. It is easy to detect the electromagnetic energy emitted by the sun. However, there is also a different type of emission from this nearby star that never reaches Earth's surface. It is a continual stream of particles and magnetic field called the **solar wind.**

Traveling at about 400.0 km/s (248.5 mi/s), the solar wind streams outward from the sun. That's almost one million miles per hour! It consists of charged particles flung by the sun's atmosphere into space. Here on Earth, we are protected by our planet's magnetic field, which acts like a shield. Particles that are moving towards Earth are mostly redirected around and then away from Earth by this field. Scientists are still researching how these particles behave. However, during higher particle events, they do know that communication systems at the Poles can be disrupted, and that radiation levels increase at higher altitudes. For this reason, at these times, many airlines will redirect scheduled flights.

Coronal Mass Ejections (CMEs)

Like Earth, the sun has a magnetic field. Extending beyond the orbit of Pluto, this powerful field affects the sun's surface. Disruptions and changes in the field can produce solar events and features such as sunspots and solar flares. They can even produce CMEs.

A CME is a magnetic bubble or twisted rope of magnetic field lines that lift off from the sun. The energetic protons are accelerated by the shockwave that forms in front of the CME as it pushes through the slower moving wind. The protons are energized and can move ahead of the actual CME. These protons can arrive at Earth in hours, but the CME can take days to arrive.

This image taken by the orbiting Solar and Heliospheric Observatory (SOHO) has been enhanced with color to help distinguish its solar features. It shows invisible electromagnetic energy emitted by the sun. In this image, you can observe an active and energetic solar surface.

In this image captured by NASA's Transition Region and Coronal Explorer (TRACE) spacecraft, we can see even more details of the sun's surface. These plumes of solar plasma are produced by disturbances in the sun's magnetic field. To get a sense of scale, the coronal loops can span 30 or more times the diameter of Earth.

In this image, violent explosions in the sun's atmosphere can be seen hurling high-energy particles into space. Along with the particles, pieces of the sun's magnetic field travel through space as part of coronal mass ejections.

As charged particles from a solar wind or CME strike Earth's magnetosphere, some are redirected along a path into the atmosphere above the Poles. An aurora is produced by electrons and protons that are present and moving within Earth's magnetosphere. They energize atoms and molecules in the atmosphere producing the lights we see in the sky.

Team Highlight
With Dr. Green in her office, Argonauts Lindsay Hannah and Joey Botros review data collected by the Space Weather Prediction Center.

22 • Operation: Infinite Potential   www.jason.org

Protecting Earth

Satellites

Dr. Janet Green is learning more about the behavior and effects of coronal mass ejections. In orbit, spacecraft and satellites are at risk. An impact with a single high-energy particle can disable a multi-million-dollar satellite. The particles can also degrade the solar cells that transform sunlight into the craft's electrical supply. Collisions with particles can also affect the satellite's onboard electronics, altering the satellite's programming. Even astronauts are at risk. Particles accelerated by CMEs can injure astronauts who are not properly sheltered.

Geostationary Operational Environmental Satellite (GOES)

NOAA's National Weather Service is one of the most advanced weather forecast systems in the world. In order to predict severe weather changes, NOAA relies on specialized satellites like GOES to provide the most up-to-date information. The GOES satellites maintain a high-altitude, geosynchronous orbit around Earth. That means they can essentially "hover" over a fixed location on Earth and provide valuable information spanning a third of Earth's surface. GOES satellites monitor and track the development of severe weather conditions such as tornadoes, flash floods, hail storms, and hurricanes. GOES satellites are equipped with space weather sensors that detect harmful radiation emitted by the sun. This information is essential for avoiding power surges on Earth, satellite failure, and radiation that can severely harm astronauts during space walks.

Type of satellite: Weather-monitoring satellite

Dimensions (main body): 2.0 m (6.6 ft) by 2.1 m (6.9 ft) by 2.3 m (7.5 ft)

Dimensions (solar array): 4.8 m (15.7 ft) by 2.7 m (8.9 ft)

Weight at liftoff: 2,105 kg (4,641 lbs)

Orbital information:

Type—Geosynchronous; Altitude—35,786 km (22,236 mi); Period—1,436 minutes (time it takes to make one orbit); Inclination—0.41 degrees (orbit is almost parallel to the Equator)

Sensors:

Data collection and relay system—ground based data communication; Imager—senses electromagnetic energy from sampled areas of Earth; Sounder—detects vertical atmospheric temperature and moisture as well as ozone distribution; Space environment/solar weather monitor—detects high energy proton and alpha particles (solar wind); Search and rescue transponder

Crew: 0

Communications

Charged particles that stream toward Earth's polar regions can also create radio interference. This can affect communication to aircraft flying at higher latitudes. To maintain communication, Dr. Green's forecast helps to warn flights so they can be re-routed. In addition, the precise signals needed for global positioning systems (GPS) can be affected.

Electrical Supplies

Changes to the planet's magnetic field can affect electrical supplies. As the field distorts, it can generate electrical energy within power lines on the surface. Surges in electricity can cause transformers, which are needed to control voltage levels, to heat up and burn out. That is what occurred on March 13, 1989 at 2:44 A.M., when most of the Canadian province of Quebec lost electricity due to a CME event.

From their station at the NOAA Space Weather Prediction Center, Janet Green and a team of specialists search for evidence of CMEs on a collision course with Earth. If a strike appears imminent, they will be in communication with individuals from the power, satellite, aircraft, communication, and military sectors. With adequate warning, they will be able to minimize damage to orbiting spacecraft, ongoing communications, and ground-based electrical grids. The data will also be shared so we may better understand how we can protect ourselves from such violent bursts of space weather.

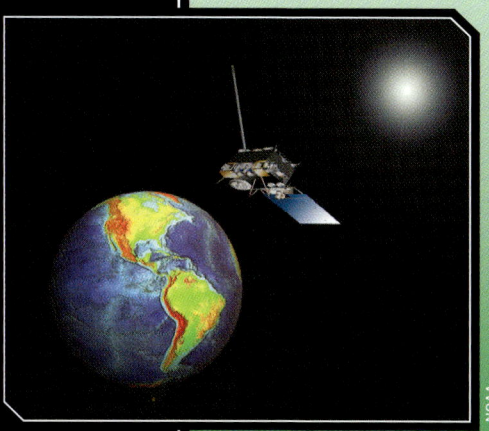

Mission 1: Critical Current—Defining Energy

Field Assignment

Exploring Energy

Recall that your mission is to *explore energy in its many forms*. Now that you have been fully briefed, you will make important space weather forecasts by analyzing actual data obtained and transmitted by space weather satellites.

The sun, in one way or another, is responsible for most of the energy we have here on Earth. It is also the main contributor to the space weather Dr. Janet Green monitors. The sun emits streams of solar wind in all different directions. Usually, this wind is relatively harmless, but if a solar flare or coronal mass ejection (CME) is released from the sun, the high-energy particles accelerated and emitted from these blasts could do some serious damage.

On October 28, 2003 at approximately 11:12 UT (Universal Time), a giant solar flare was observed using an ultraviolet imaging telescope on one of NASA's space weather satellites. The flare produced x-rays (electromagnetic energy) that blasted towards Earth at the speed of light (2.99 x 10^5 km/s). Space weather scientists closely monitored the sun for further activity.

Approximately 18 minutes later, space weather satellites observed a CME. This CME accelerated protons towards Earth at speeds reaching 90,000 km/s. Following these protons was the CME, a huge bubble of gas threaded with magnetic field lines, hurtling towards Earth at approximately 2,300 km/s.

In this field assignment, you will analyze data collected from space weather satellites on October 28, 2003. You will use mathematical equations to help forecast when x-rays, protons and the CME will make contact with Earth. You will assess the accuracy of your prediction by comparing it to the actual contact time. Once you have developed skills in space weather forecasting, you will use your knowledge of space weather and energy forms to survey and evaluate the strengths and limitations of energy forms in your local area.

Objectives: To complete your mission, accomplish the following objectives:

- Analyze weather satellite data from a recent CME.
- Apply mathematical equations to forecast when high-energy particles will collide with Earth.
- Compare your prediction with the actual impact time.
- Identify people or organizations to contact in order to help prepare and protect Earth when a CME blasts towards Earth.
- Survey and document a variety of different forms of energy in your local area.
- Evaluate strengths and limitations of the different forms of energy in your local area.

Materials
- Mission 1 Field Assignment Data Sheet
- digital camera
- math tools (p.114)

 Caution! You must have the landowner's permission to access and photograph energy forms in your study. Never travel alone and take a responsible adult with you to the study sites at all times.

Field Preparation

1. Identify and describe the consequences associated with x-rays, CMEs, and protons accelerated by CMEs when they collide with satellites, space stations, and Earth. Identify organizations that should be contacted in case of a CME.

2. Understand how to apply and manipulate the equation:

$$\text{average speed} = \frac{\text{total distance}}{\text{total time}}$$

3. Analyze data collected by NASA satellites and use mathematical calculations to predict how much time you will have to contact organizations that would be affected by these x-rays, protons, and CME.

4. Compare your results with the actual time the x-rays, protons, and CME were detected by Earth-orbiting NASA satellites.

Mission Challenge

1. With a responsible adult, survey and document (photograph or draw) a variety of energy forms in your local area.

2. Create a visual presentation which:

   a. Identifies and describes the forms of energy in your local area.

   b. Classifies these energy forms as potential energy or kinetic energy.

   c. Explains how these forms of energy affect your everyday life.

   d. Describes how a CME can affect these energy forms, and if so, explains the consequences.

   e. Assesses the strengths and limitations of these energy forms.

Mission Debrief

1. Describe how we use energy to make our lives more comfortable.

2. If humans did not have the ability to harness and use energy, what do you think your life would be like today? Explain.

 Journal Question Describe how we know energy is present, even though it is not always visible.

Mission 1: Critical Current—Defining Energy

Connections

History

Sun Worship

It is May 26, 1954. Lowering his head into the gap, Kamal el-Mallakh digs into the surrounding rock. As the Egyptian archeologist cuts deeper into the limestone, the opening widens. Rock crumbles as he burrows into a hidden underground chamber. In the stale air, he recognizes the distinct smell of cedar wood, a scent trapped within the sealed chamber for over 4,000 years. Success! El-Mallakh has uncovered the solar barge of the great Egyptian Pharaoh Khufu, also known to the Greeks as Cheops.

Unlike other craft of its time, el-Mallakh's find was not built to cruise on the Nile River carrying wheat or flax. Its cargo was far more precious. The barge's role was to carry the soul of the deceased pharaoh into the afterlife where he would join his father, the sun god, Ra.

The Ancient Egyptian civilization rose up along the banks of the Nile as an agricultural society. Because their crops depended on sunlight for a successful harvest, its residents took a great interest in the sun. They tracked it over days, months, and years. By recording its cycles, they could better understand its behavior and relationship to the plants they harvested.

To ancient Egyptians, the sun traveled across the sky along with various other gods in a vessel called a solar barge. At night, the barge would sink into the underworld. There, it would reverse direction and flow from west to east. At dawn, the sun barge would once again rise in the east and follow its customary track across the sky.

Have you ever seen a hieroglyphic? It is an ancient way of writing that uses pictures and letters to record and communicate ideas. Many hieroglyphics show gods and god-

▲ Much of our knowledge of ancient Egyptian civilization is based upon interpretations of hieroglyphics and Egyptian artwork captured on papyrus. Many of these works incorporate symbolism expressed through the use of specific colors, animals, and objects.

26 • Operation: Infinite Potential   www.jason.org

Stone of the Sun

The Stone of the Sun is a historic sculpture that measures about 3.6 m (11.8 ft.) in diameter. It was carved in the fifteenth century and later buried. On December 17, 1790, it was unearthed in the main square of Mexico City. The face in the center of the stone is most likely that of Tonatiuh, the fifth sun god. Surrounding him are four boxes that represent each of the previous suns and worlds that were destroyed. The Stone of the Sun is now on display at the National Museum of Anthropology in Mexico City, Mexico.

Earth's continued existence was not guaranteed. According to legend, it had been destroyed four times by the death of the sun. To prevent its fifth destruction, they needed to nourish Tonatiuh, a sun god who battled the forces of darkness.

The Aztec sun god was fortified by human sacrifice. Although the numbers are debated, most historians believe that for many years, thousands of sacrifices were made to the sun god. Often, the heart would be removed from the victim while still beating, and placed in a sacrificial bowl for the god.

desses carrying the sun on their heads. This demonstrated the society's close association with Ra, who was a symbol of power.

Throughout history, many other cultures developed beliefs that included a sun god. In Mesopotamia, the sun deity was called Shamash. In Greek mythology, he was called Helios or Titan. The Roman myths identified him as Sol Invictus, which meant *unconquered sun*. In other interpretations, this sun god was actually a goddess. Like her male counterparts, she conveyed both the life-giving aspects and the awesome power of this brilliant orb.

In a region of central Mexico, the ancient Aztec civilization practiced a much more brutal form of sun worship. To the Aztecs,

YOUR TURN

Use online and print resources to learn about an ancient society or tribe that worshipped the sun. Present your findings to classmates using models and drawings that you have made to convey various aspects of this worship.

Connections: History • 27

Mission 2
Waves of Change
Calculating Transfers and Transformations

"In one instant, this terrible disaster has done more to raise awareness of the tsunami threat than anything we have tried over the years."
—Dr. Vasily Titov
Director, NOAA Center for Tsunami Research

Vasily Titov

Vasily Titov and his research team use Deep-ocean Assessment and Reporting of Tsunamis (DART®) buoys, tide gauges, seismic sensors, and computers to monitor ocean activity in real-time. He hopes to develop a system that will forecast and warn coastal communities of possible tsunami threats before they strike.

Meet the Researchers Video
Discover how tsunamis are detected from the ocean bottom and how researchers like Vasily Titov work together to help protect and guard our coastal shores.

Director, NOAA Center for Tsunami Research

Read more about Vasily online in the JASON Mission Center.

Photos above (left to right): T. Ewy/Wikimedia Commons; NASA; Whitney Caldwell, The JASON Project; Darren Hester/OpenPhoto.net-SA by 2.0; NASA; David Rydevik/Wikimedia Commons; Rube Goldberg is the ® and © of Rube Goldberg, Inc.; Wikimedia Commons; Wikimedia Commons.

Your Mission...

Understand how energy is transferred and transformed in predictable ways.

To accomplish your mission successfully, you will need to

- Analyze energy transfers and transformations within a system.
- Explore the properties and behaviors of waves.
- Describe how waves transfer energy.
- Apply an understanding of waves to tsunami formation.
- Investigate energy efficiency.

Mystery Connection

ALERT!
To: JASON Tsunami Warning Center
From: NOAA
Subject: Tsunami Information Needed

Preliminary earthquake parameters:

Magnitude	8.7 on the Richter scale
Time	11:44 HADT (local time)
Location	17° south 173° west
Depth	86 km (53 mi)
Plate movement	The Pacific plate is being subducted under the Australian plate at this location, called the Tonga Trench.

Based on the incoming data, should we evacuate or warn any communities of a possible tsunami? If so, who, where, and how?

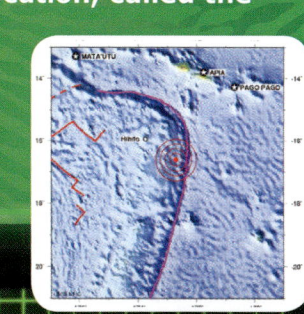

Mission 2: Waves of Change—Calculating Transfers and Transformations • 29

Guarding Our Shores

December 26, 2004, 7:58 a.m. North Sumatra, Indonesia. A cool breeze blows in from the Indian Ocean. The city streets are filled with morning traffic. Stores and shops open their doors as people get ready for the day to begin. Unaware of the danger that lurks in the nearby ocean, the residents of Sumatra go about their daily lives.

Halfway around the world in Seattle, Washington, Dr. Vasily Titov is finishing his dinner. Checking the incoming message on his beeper, he grabs his jacket and runs out the door. Racing through the city, he knows he must act quickly. With no time to lose, he weaves through traffic while completing some preliminary calculations in his head.

Two hours after an earthquake has rocked the Indian Ocean, he sits down at his computer. As he analyzes the incoming data, an image representing a 1,600 km-long (1,000 mi) tsunami screams across his screen as his simulation begins. By 4:30 a.m. in Seattle, his simulation has run its course, but many shores along the Indian Ocean are already devastated.

These are true events. Although the final death toll still remains unknown, many believe that the tsunami killed over 300,000 people in the Indian Ocean basin. The impact of this single event has been a major force in accelerating the implementation of a tsunami forecasting system. Dr. Vasily Titov and his team of researchers are at the forefront of this effort. Stationed at Pacific Marine Environmental Laboratory (PMEL) in Seattle, these scientists monitor data from an international network of 46 Deep-ocean Assessment and Reporting of Tsunamis (DART®) buoys. Positioned in the Indian, Pacific, and Atlantic Oceans, these buoys transmit real-time data that can be used to forecast incoming tsunamis before they strike.

 Mission 2 Briefing Video Prepare for your mission by viewing this briefing on your objectives. Learn how Dr. Vasily Titov uses his knowledge of energy transfer and transformation to help save lives and protect property.

Fast Fact
The highest tsunami on record happened in 1958 in Lituya Bay, an inlet of Glacier Bay National Park in Alaska. A nearby earthquake triggered a landslide that tumbled into the bay. The impact produced a wave that was estimated to be the height of a ten-story building. As this wave struck the opposite shoreline, water splashed upward to a reported height of over 500 m (1,640 ft).

Mission Briefing

Systems

As you strap yourself into the seat and pull down the shoulder harness, you feel your heart beat a little faster. With the push of a button, the motor comes to life and the roller coaster car starts its slow journey up the first hill.

When you ride a roller coaster, you become part of a **system**. Like all systems, the roller coaster is composed of a related set of components. Some of the components are physical objects, such as the car and track. Other parts of this system are processes, such as the cars' movement and the related energy transfers and transformations.

Energy Transfers

In a system, **energy transfers** take place when energy stays in the same form but is passed between different objects. Think of a row of dominoes. The mechanical energy of the first domino is transferred to the mechanical energy of the next domino and so on. The energy remains mechanical as it is passed on to different objects.

30 • Operation: Infinite Potential   www.jason.org

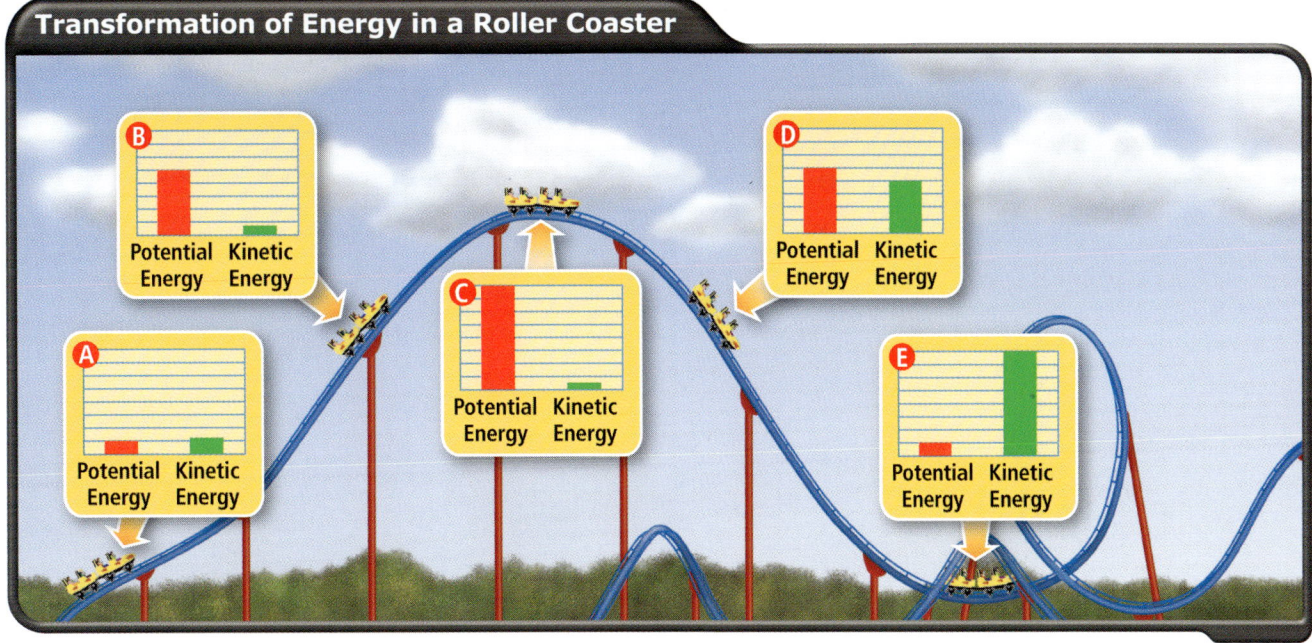

Transformation of Energy in a Roller Coaster

Energy Transformations

Energy transformations occur in a system when energy changes into different forms. Energy transformations can occur within an object, or between objects. Think about the engine in a car. The chemical energy of the fuel is transformed into the thermal and mechanical energy of the motor and tires.

Dr. Titov studies a system that involves energy transfers and transformations. However, instead of studying amusement park rides or dominoes, he investigates the natural system of energy transfers and transformations that can produce tsunamis.

Tsunamis are a special type of water waves. Waves you see at the beach are usually created when wind transfers energy to the ocean's surface. However, with tsunamis, the potential energy in tectonic plates or an underwater landslide can be transformed into the kinetic energy of moving tsunami waves.

Join the Team

At the Pacific Marine Environmental Laboratory in Seattle, WA; Tim West, Bryan Ie, Toba Faseru, Madhu Ramankutty, Dr. Vasily Titov, and Chris Meinig (L to R) investigate the properties of waves and how the transfers and transformations of energy from a disturbance on the ocean floor can lead to a destructive tsunami wave. Dr. Titov uses this information to create models that may predict where and when a tsunami will strike land, to help warn coastal residents of impending danger.

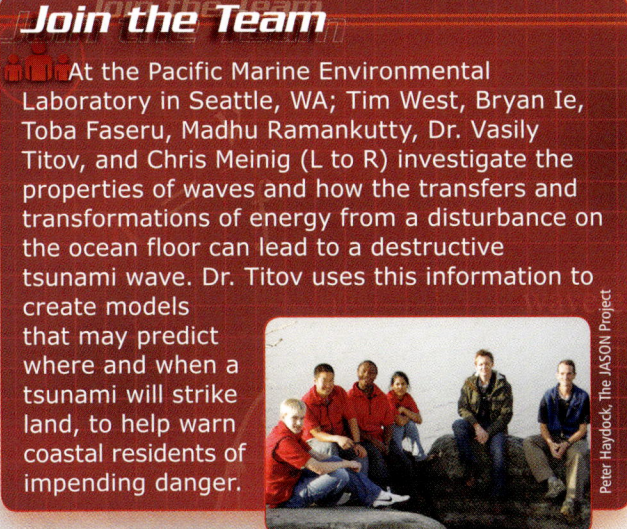

Example

As a roller coaster car climbs the first hill, kinetic energy **transforms** into potential energy. Then, as the car races down the first hill, the potential energy in the car is rapidly transformed back into kinetic energy. Roller coaster engineers and designers know that the **transfers** and **transformations** are not that simple. As the wheels get hotter and the car makes noise, energy transformations remove energy from the car to other parts of the system. These transformations cause the car to slow down.

Conservation of Energy

Even though a roller coaster car slows down as it moves along the track, we know that the energy is not being destroyed. In fact, energy is neither created nor destroyed within a system; it is only transferred or transformed. As the roller coaster car races down each hill, gravitational potential energy is transformed into mechanical, sound, and thermal kinetic energy. This law of **Conservation of Energy** helps engineers design roller coasters, and helps scientists forecast tsunamis and even study the solar system.

Expanding upon the law of Conservation of Energy, Albert Einstein established that matter can be considered a form of energy. We see this in nuclear reactions where some matter is transformed into thermal and electromagnetic energy.

Mission 2: Waves of Change—Calculating Transfers and Transformations

Lab 1

Energy Transfers and Transformations

To forecast tsunamis, Dr. Vasily Titov must understand the complex system of energy transfers and transformations that result in these events. Although a humorous cartoon may seem very different, the artist Rube Goldberg is known for creating complicated machines that also involve energy transfers and transformations.

In this activity, you will have the opportunity to analyze one of Goldberg's cartoons and create your own multi-step machine that is based upon energy transfers and transformations. Then, you will apply these concepts to create a contraption that requires several energy transfers and transformations to complete a task.

Self-Operating Napkin

Materials
- Lab 1 Data Sheet
- materials provided by instructor

Lab Prep

1. Analyze the cartoon. What task is eventually completed? What is the initial action that begins the sequence? How many distinct steps can you identify in the machine's mechanics?

2. What is the difference between energy transfer and energy transformation?

3. Make a list of all the energy transfers and energy transformations shown in this machine. What other transfers and transformations are assumed, but not incorporated into the illustrated steps of the machine's actions? Explain.

Make Observations

1. Examine the materials supplied by your instructor. Brainstorm a sequence of energy transfers and transformations that could be assembled with these items to perform a task. Include at least two energy transformations.

2. Draw a blueprint of your proposed Rube Goldberg machine that includes a description of energy transfers and transformations. Share your design with the instructor and with your instructor's approval, proceed with the assembly of your machine.

3. Once the assembly is complete, test your machine. Does it work? Could it be improved? If so, how?

4. Without any material constraints, design a more complicated machine that can complete a different task using as many energy transfers and transformations as possible. What task will your machine complete? How will the sequence begin? How will it end?

5. Create a blueprint of your new Rube Goldberg machine, identifying and describing all the energy transfers and transformations. Share and discuss your machine with members of your class.

 Journal Question Does the machine you created in Make Observations step 2 illustrate the Conservation of Energy? Explain.

Waves

Have you ever seen lightning strike and then heard thunder a few seconds later? Energy does not transfer instantly. It takes time to move from one place to another. Electromagnetic energy transfers faster than sound energy, which is why you can see the light of lightning before you hear the sound of thunder. Electromagnetic energy and sound both travel as waves. **Waves** are the progressive disturbances that transfer energy from one place to another.

Mechanical and Electromagnetic Waves

Waves can be broken down into two main categories: mechanical and electromagnetic. **Mechanical waves**, such as sound or tsunamis, must travel through a medium such as air or water. When there is no medium, like in space, mechanical waves cannot exist. Mechanical waves travel at different speeds as they move through solids, liquids, or gases.

Even though different types of mechanical waves travel at different speeds, they share many properties. All mechanical waves start with a disturbance. Strum a guitar string or drop a stone in a pond and you create a disturbance.

Only the energy of the disturbance travels through a medium. The medium itself is not carried along with the wave. Think of the "wave" sports fans sometimes do in crowded stadiums. People only move up and down as the "wave" moves through them. The individual person is not carried along with the wave, but remains at her seat.

All parts of the electromagnetic spectrum, from radio waves to gamma rays, use electromagnetic waves to transfer their energy. **Electromagnetic waves** do not require a medium through which to travel. These waves can exist in space. When electromagnetic waves travel in space, they travel at or near the speed of light, which is about 300,000 km/s (186,000 mi/s) in a vacuum.

Wave Movement

When energy moves in waves, it can be observed as either compression waves or transverse waves.

Compression Waves

Compression waves vibrate the medium back and forth in the same direction that the wave travels. You can make a compression wave by squeezing together and releasing several coils of a stretched spring. Repeat this process and you will notice that some areas become squashed together while other areas get stretched out.

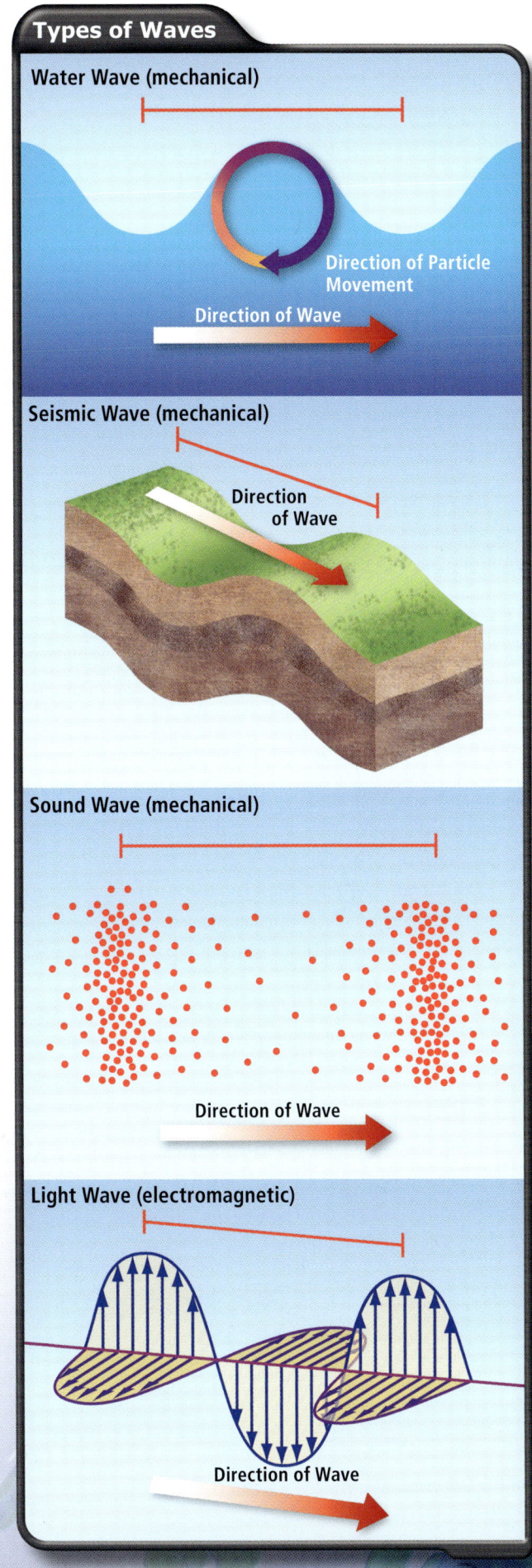

Mission 2: Waves of Change—Calculating Transfers and Transformations • 33

The areas that are squashed together are called **compressions**. On either side of the compressions, the spring coil is being stretched. These areas are called **rarefactions**.

Sound waves move as compression waves. Hold your hand in front of a speaker playing music and you can feel these compression waves. When a speaker moves in and out, it transfers kinetic energy to the air particles next to it. The speaker's movement pushes and pulls the air particles, creating compressions and rarefactions in the air. Compression waves, such as sound waves, can use solids, liquids, or gases as a medium. However, the speed of the wave is affected by the medium it is traveling within.

Transverse Waves

Transverse waves move at right angles, or perpendicularly, to the direction the energy travels. You can see transverse waves when you shake one end of a jump rope from side to side or up and down. While the energy moves along the rope from end to end, the disturbance that created the wave moves at right angles to the direction of the energy. Electromagnetic waves are also examples of transverse waves.

Describing Waves

Observe waves at the beach and you will notice the basic parts of a wave. The water waves you can see are transverse waves, which have a distinct crest and trough. **Crests** are the high points of the wave, and **troughs** are the low points between crests. The height of each crest or the depth of each trough from the center is called the **amplitude**.

Example

The speed of **sound waves** depends on the medium and temperature. These waves travel faster through solids, whose particles are closer together, compared with liquids and gases, whose molecules are farther apart. For example, sound travels around 12,000 m/s (39,000 ft/s) through diamond (a solid) and only 343 m/s (1,130 ft/s) through air (a gas).

In addition to the type of medium, the speed of a compression wave depends upon temperature. For example, sound travels around 331 m/s (1,090 ft/s) through air at 0°C (32°F) and increases to 343 m/s (1,130 ft/s) through air at 20°C (68°F).

In a compression wave, the compressions are similar to the crests of a transverse wave and rarefactions are similar to the troughs. The amplitude of a compression wave is measured by the density of the medium at its greatest compression.

The distance from crest to crest or trough to trough in a transverse wave is called the **wavelength**. In a compression wave, wavelength is measured from compression to compression or rarefaction to rarefaction.

Suppose you stood in the water and counted the number of wave crests that passed by you in a given amount of time. This count would tell you the **frequency** of these waves. In a compression wave, this frequency would be measured by determining the number of compressions over time.

Transverse Wave
crest
trough

Compression Wave
rarefaction
compression   compression

Team Highlight
Bryan Ie, Toba Faseru, Tim West and Madhu Ramankutty (L to R) attach a cover to the buoy that Dr. Titov and his team will deploy to the Pacific Ocean. This buoy relays information from another device at the bottom of the ocean floor, called a tsunameter, which senses when a possible tsunami wave has passed by.

Whitney Caldwell, The JASON Project

Tsunami Generation

1. The flow of molten material beneath the plates produces a force that causes plate movement.
2. At some boundaries, the plates bind together and so the kinetic energy that would have moved the plates becomes stored as potential energy.
3. Suddenly the stored energy is released. The potential energy transforms back into kinetic energy. This is an earthquake.
4. The kinetic energy of the moving plates is transferred to the water, sometimes creating a tsunami.

→ Kinetic Energy
○ Potential Energy

Tsunami Wave Formation

The energy required to create tsunami waves is most commonly generated by underwater earthquakes when tectonic plates suddenly slip along plate boundaries. As tectonic plates push together, the kinetic energy of the moving plates is transformed into potential energy. This potential energy is stored at the plate boundary.

Over time, the potential energy increases so much that it overcomes the friction between the plates, which has kept them motionless. The plates can very suddenly and violently slide past one another creating earthquakes. As the plates move vertically, they transform the built-up potential energy back into the kinetic energy of movement. The upward movement of the seafloor transfers kinetic energy from the solid earth to the ocean water above. This then gives the water kinetic energy, moving water upwards. When this vertical movement reaches the ocean surface, it can be transformed into waves. These waves begin moving horizontally outward toward shorelines as a tsunami!

Mission 2: Waves of Change—Calculating Transfers and Transformations • **35**

In the deep ocean, a portion of the wave energy is kinetic. Tsunamis move up to 800 km/h (550 mph). The amplitude of these waves is very low. This is why tsunami waves can go right under boats in the open ocean without even being noticed by people on board.

As tsunami waves approach the shore, the seafloor causes the waves to slow down. This decrease in speed causes the waves to get closer together and taller. The waves now have a shorter wavelength and greater amplitude. As the wave height increases, the kinetic energy of the water is transformed into potential energy. Eventually, the tsunami washes onto the shore, potentially flooding low-lying areas with the strength to destroy buildings, or even causing lives to be lost.

Predicting Tsunamis

Dr. Titov has dedicated much of his career to developing a system to forecast tsunamis. As a mathematician, he has helped write computer programs with geologists, oceanographers, and other mathematicians which model tsunami formation. He uses data from tsunami events in the past to improve the models.

Vasily also collects data as earthquakes happen around the world. This data, in combination with data collected from DART® buoys, is entered into the tsunami forecast programs that highlight threatened coastal areas. Each time there is an earthquake or landslide in the ocean, he and his team are able to refine the forecasts made by the computer programs.

Energy in Wave Movement

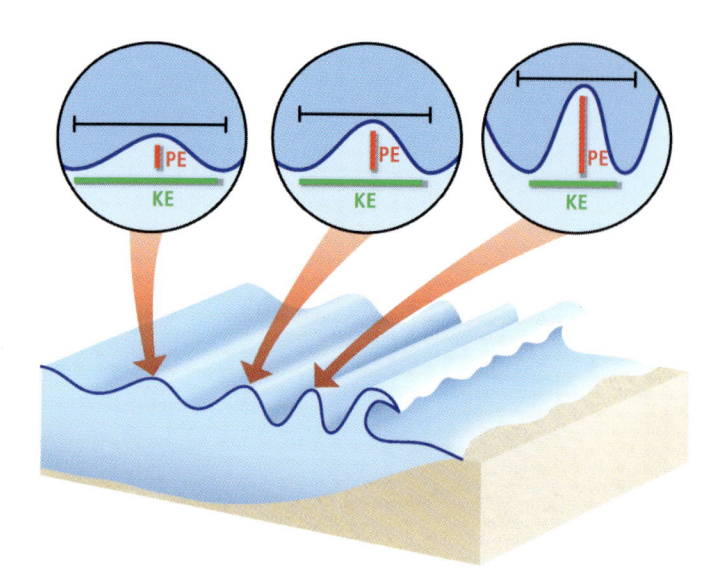

▲ As waves move from the deep ocean toward the shore, their wavelengths decrease and amplitudes increase. The kinetic energy of the water is transformed to potential energy, as seen in the height of the wave.

As more DART® buoys are deployed and faster computers are used, this will help refine the team's forecast programs. Dr. Titov hopes this will result in a quicker and more accurate warning system that will someday be powerful enough to save every person in the path of a tsunami.

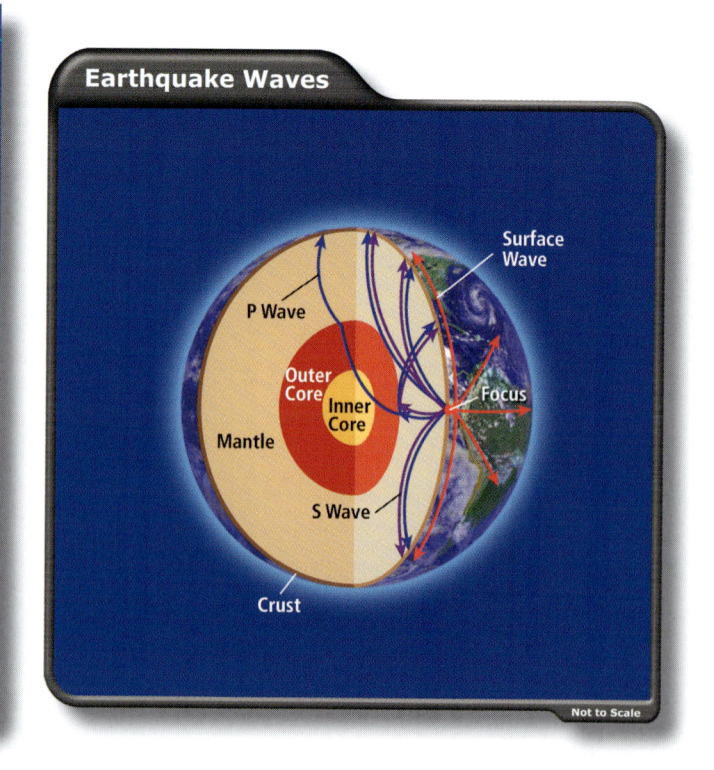

Fast Fact

Most of the data scientists have gathered about the interior of Earth has been collected using our understanding of earthquake waves. Mechanical waves, such as earthquake waves, are affected by the medium through which they travel, and their behavior changes as they move between different mediums. Scientists have found that the P and S waves of an earthquake do not move in a straight line through Earth. Rather, they reflect and refract as they travel through the interior. If no boundaries between the layers of Earth existed, we would not detect any changes in wave direction. This behavior has led scientists to develop a model of Earth's interior as composed of layers of many different materials.

Earthquake Waves

36 • Operation: Infinite Potential   www.jason.org

Lab 2

Wave Tank Tsunami

Like other scientists, Dr. Titov uses models to better understand the processes that generate tsunamis. Most of his models are computer simulations that use real-world data to make predictions about tsunami formation.

In this activity, you will have an opportunity to observe the behaviors and properties of waves. Based upon your observations, you will understand how the devastation produced by this wave is affected by shoreline characteristics.

Materials
- Lab 2 Data Sheet
- clear baking dish
- dropper
- 2 block magnets
- waterproof clay
- sheet of white paper
- flashlight
- wooden blocks (or alternate supports)

▲ Overhead observations reveal how waves are affected as they approach and interact with coastlines.

Lab Prep

1. Use wooden blocks or other supports to elevate the dish at least 10 cm above a piece of paper on your desktop.

2. Add water to the dish. Stop when the dish is filled to about 3/4 of its depth.

3. Blow across the water in short puffs and longer sustained breaths. What do you observe? What process is modeled by blowing on the water?

4. Release one drop of water from a dropper into the center of the dish from a height of at least 25 cm. Describe what you observe.

5. Aim a flashlight beam directly down at the point where the drop strikes the water's surface. Release another drop. Describe what you observe on the paper below the dish.

6. Release a drop closer to the side of the dish. Describe what happens now. Explain any changes you see.

7. Place pieces of clay near the center of the dish, making sure they extend above the water's surface. How does the clay's presence affect the pattern of waves produced by the falling droplets? Make a diagram of what you observe. Can you build a barrier that restricts the spreading waves?

Make Observations

1. Remove the clay from your dish.

2. Press a lump of clay onto the top of a block magnet, making sure that it remains affixed to the metal. Position the magnet underwater at one end of the dish. Make sure the clay does not project above the water's surface.

3. Place the second magnet on the outside of the dish beneath the submerged magnet. Make sure that the magnets are strong enough to move each other through the thickness of the dish bottom.

4. Move the outer magnet. How does the submerged magnet respond? Explain your observations. What geologic process is modeled by the magnet's movement?

5. Does the movement of the submerged magnet affect the water's surface? How?

6. Use clay to mold a coastline with a large central bay at one end of the dish. Predict how the shape of the model coastline will affect the wave as it moves into the bay. Test your prediction.

Journal Question Describe two ways in which ocean waves can be produced.

Mission 2: Waves of Change—Calculating Transfers and Transformations

Thermal Energy

Heat and Temperature

The words *hot* and *cold* refer to how we perceive an object's thermal energy. **Thermal energy** is the total energy content of a system. Often, it is measured in terms of particle vibration and movement. Particles that move or vibrate faster have greater thermal energy than particles that move at slower speeds.

Temperature is a measure of the concentration of thermal energy, not the total amount of thermal energy in a system. Although the terms are often incorrectly interchanged, do not confuse thermal energy with temperature.

Like other types of energy, thermal energy can be transferred and transformed. In all cases, thermal energy moves from areas of high concentration to areas of low concentration. This flow of thermal energy between substances is often referred to as **heat**.

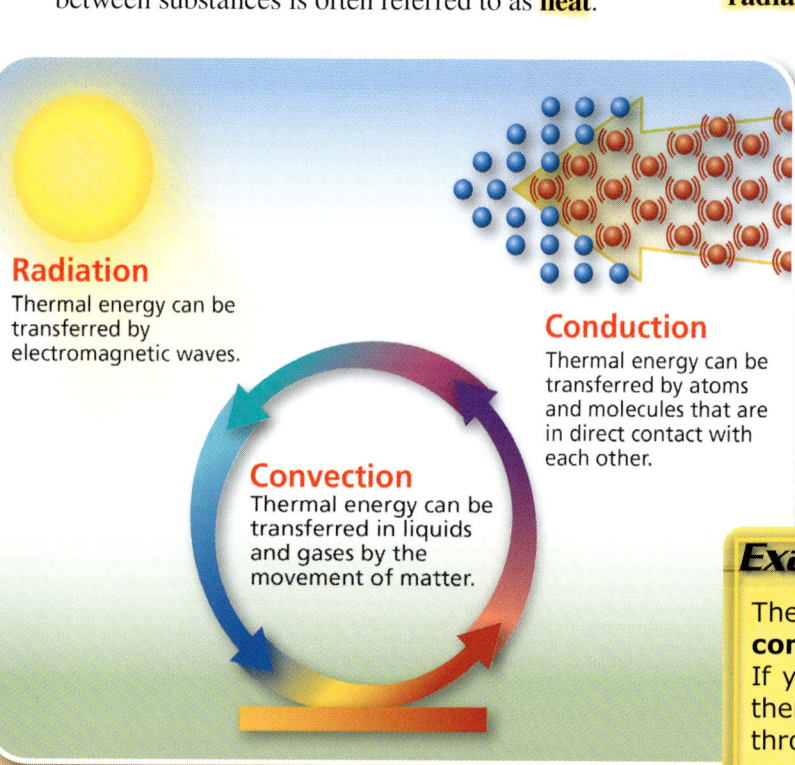

Fast Fact

Compare a mug of hot tea and a swimming pool. The tea's temperature is 82°C (180°F). The swimming pool's temperature is 25°C (77°F). Although the measured temperature of the tea is higher, the pool's total content of thermal energy is greater. That is because there are many more moving particles in the pool. Even though the particles in the pool are slower moving, their numbers account for a much greater amount of thermal energy.

Conduction, Convection and Radiation

Thermal energy is transferred or transformed between substances by **conduction**, **convection**, or **radiation**. The process that transfers thermal energy between atoms and molecules that are in direct contact is called conduction.

Convection transfers thermal energy through the movement of matter and occurs in materials that are capable of flowing, such as liquids and gases. Colder liquids and gases are more dense and therefore fall due to gravity, displacing warmer material, which rises. This movement forms a convection cycle.

Radiation is the transfer of thermal energy through electromagnetic waves. The flow of energy from the sun to the planets is an example of radiation.

Radiation
Thermal energy can be transferred by electromagnetic waves.

Conduction
Thermal energy can be transferred by atoms and molecules that are in direct contact with each other.

Convection
Thermal energy can be transferred in liquids and gases by the movement of matter.

Example

Thermal energy can be transferred by **conduction**, **convection**, and **radiation**. If you put a spoon in hot soup, over time, thermal energy transfers from the hot soup, through the spoon to your hand. This is **conduction**.

Convection within Earth transfers energy to the surface. These transfers drive geological processes that can result in earthquakes and tsunamis. As magma circulates, thermal energy is transferred from the core to the surface.

If you have ever stood in the sun on a hot summer day, you have felt the thermal energy transferred from the sun to your skin by **radiation**.

Lab 3

Thermal Energy Survey Lab

Thermal energy from Earth's mantle drives the massive movements of undersea crust, which result in seismic events that can potentially generate the tsunamis studied by Dr. Titov. Thermal energy is the form of energy that we associate with the vibration and movement of atomic particles. As these bits of matter gain thermal energy, their movements increase. As you have observed, these changes in motion affect various properties of the material, including its state, volume, and rate of chemical reactivity.

In this activity, you will have the opportunity to explore thermal energy. You will use tools to observe its presence and effects. You will also construct an understanding of this energy through several inquiry-based investigations.

Materials
- Lab 3 Data Sheet
- **convection detector tool** (p.114)
- safety goggles
- materials provided by instructor

Lab Prep

1. When you enter the laboratory, do not touch any of the stations. Wait for instructions from your instructor before beginning the investigation.

2. With your instructor, review all appropriate laboratory procedures, safety guidelines, and classroom rules.

3. Review the objectives and any procedures that are established for each station.

4. Review the order in which you will move from station to station.

Make Observations

1. After your instructor has presented the lab prep, a signal will be given to begin work at your first station. Remember that you have a limited amount of time to work at each station and answer questions.

2. Use the tools and materials that are available at that station to perform the investigations.

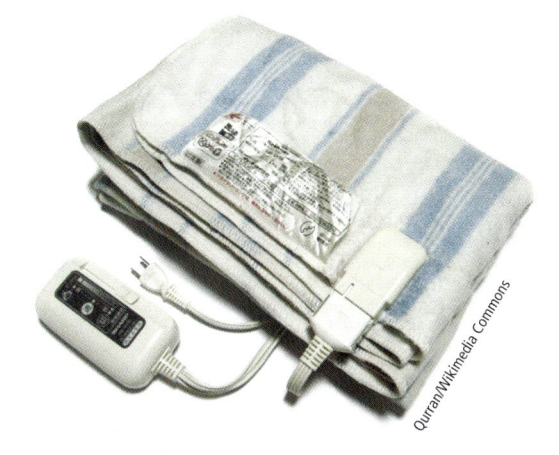

Journal Question Identify and describe five different sources and uses of thermal energy from your everyday experiences.

Team Highlight

Madhu Ramankutty and Teacher Argonaut Bryan Ie explore how energy is transferred and transformed by creating a sequence of events to complete a task.

Mission 2: Waves of Change—Calculating Transfers and Transformations • 39

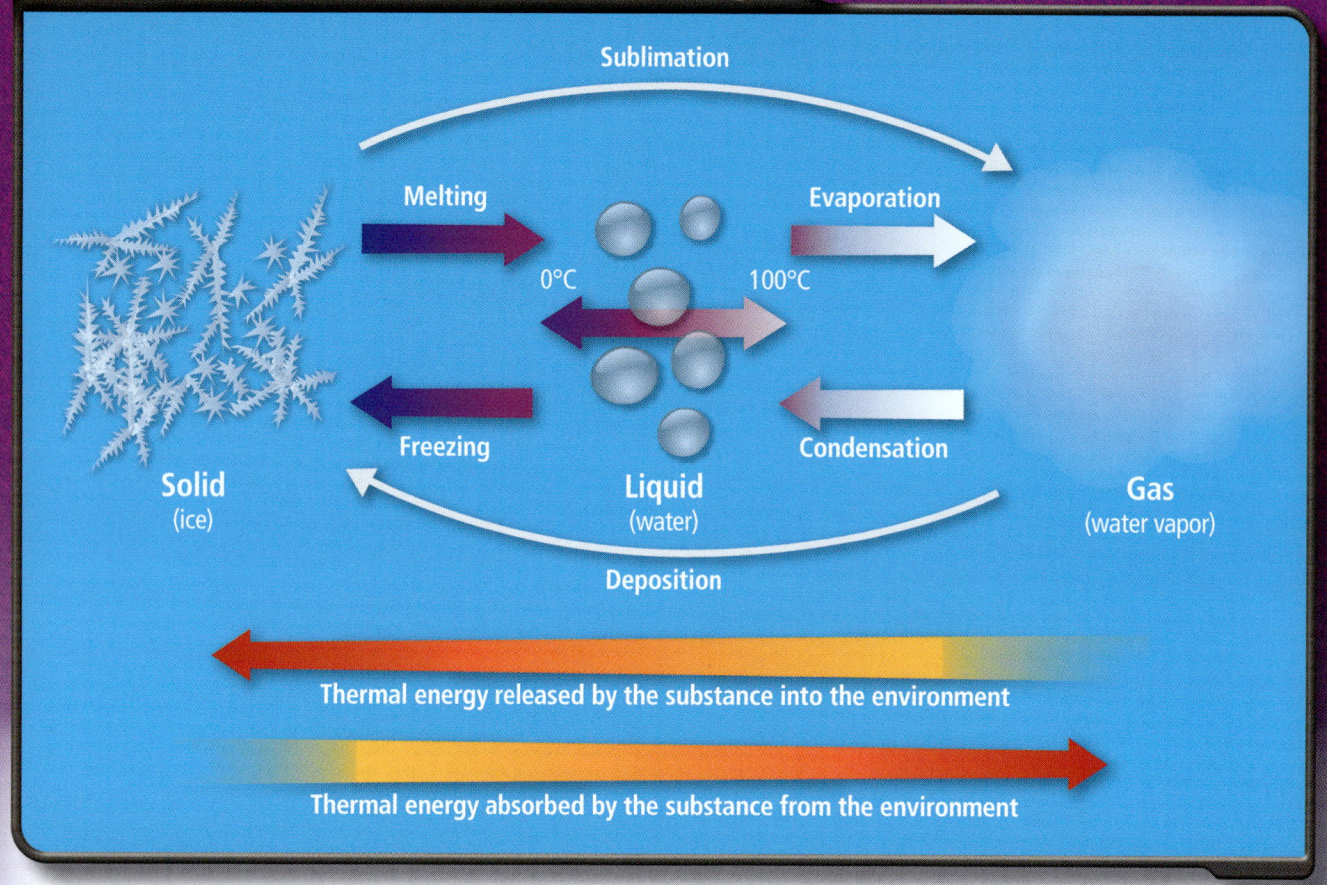

Transfer of Thermal Energy During Phase Change

Sublimation

Melting → **Evaporation** →

0°C 100°C

← **Freezing** ← **Condensation**

Solid (ice)   **Liquid** (water)   **Gas** (water vapor)

Deposition

← Thermal energy released by the substance into the environment

Thermal energy absorbed by the substance from the environment →

Conductors and Insulators

Thermal energy transfers within a system are regulated by the property of conduction. Good **conductors** are substances that easily transfer energy. The metal of a soda can is a good conductor. It feels cold because the thermal energy easily transfers from your hand to the can.

Insulators do not easily conduct energy. A plastic cooler is a thermal insulator. This property means that thermal energy cannot be easily transferred into or out of the cooler.

Cold

Cold is a word used to describe a condition of low temperature. It is critical when talking about cooling and refrigeration. Cooling objects with refrigeration involves putting them into an unbalanced system. The unbalanced system promotes a transfer of thermal energy away from the substances you want to cool.

Air conditioners also create this imbalance. When you enter an air-conditioned room, your temperature is immediately out of balance with the room. This causes your body to release its thermal energy to the surroundings. The constant loss of your thermal energy to the surrounding air causes you to feel cooler.

Example

Your body has its own **cooling system**. Along with many other mammals, you have glands all over your body that produce sweat on your skin when you get hot. As this liquid covers your skin, it starts evaporating into the surrounding air. When substances change phase from liquid to gas, they absorb thermal energy from their local environment. Because your skin is part of the local environment, thermal energy is removed from your skin, making it cooler.

40 • Operation: Infinite Potential www.jason.org

Lab 4

Electrochemical Cells

Dr. Titov monitors ocean activity using information collected by tsunameters located on the ocean floor. DART® buoys receive information from tsunameters and transmit this data to satellites using specialized onboard electronics. Electronic transmitting and receiving devices on both the DART® buoys and tsunameters are powered by D-cell battery packs.

Each battery pack contains a set of electrochemical cells. Like household batteries, these cells store chemical energy. As the cells' electrodes react, an electrical current is produced. This flow of electrons travels through an external circuit that powers the connected electronic device.

In this activity, you will construct and explore the properties of a basic electrochemical cell. Then, you will assemble a battery pack using these cells and use the pack to power an LED.

Materials
- Lab 4 Data Sheet
- 4 iron washers
- 4 copper pennies
- 4 pieces of potato
- 5 connecting wires (with alligator clips)
- multimeter
- low-power LED

Lab Prep

1. Insert an iron washer halfway into a small piece of potato.
2. Insert a copper penny into that same potato piece. Make sure that the penny and washer are separated by several centimeters.
3. Attach one lead from the multimeter to the washer. Attach the other lead from the multimeter to the penny. Set the multimeter to DC voltage detection. What do you observe? Would changing the leads affect the measurement? Test your prediction.

Make Observations

1. Repeat steps 1 and 2 from Lab Prep to produce a second copper-iron cell. Measure the voltage of this single cell. How does it compare to the first cell you assembled? Explain and explore any difference in measurements.
2. Use one wire with alligator clips to connect the two cells. The wire should be attached to the washer of one cell and the penny of the other. Once the cells are wired together, attach the multimeter leads to the unconnected penny of one cell and the unconnected washer of the other. Record and explain your observations.
3. Suppose you were to keep on adding additional cells. How might each additional cell affect the generated voltage? Create a hypothesis. Then, test your predictions by adding two more cells to the setup.
4. Remove the multimeter from the circuit. Wire in a low-power LED in the place of the multimeter. Did you generate enough voltage to light up this device? If not, add additional cells to the circuit. HINT: The LED needs to be connected with the correct polarity. If it does not illuminate, try switching the connection wires.

Journal Questions Identify the different types of batteries you are aware of and explore the differences and variety of uses.

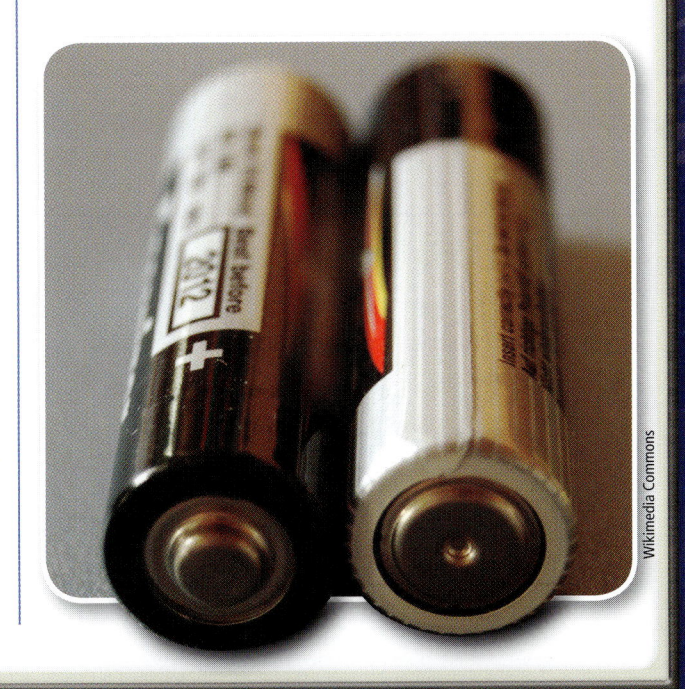

Mission 2: Waves of Change—Calculating Transfers and Transformations

Efficiency in Energy Transfers and Transformations

Efficiency

When energy transfers and transforms, some of that energy dissipates into forms that were not intended. For example, in a car, the chemical energy of the burning fuel is eventually transformed into the mechanical energy of the moving car. However, not all of the energy transforms into movement—a large portion of the energy is transformed into thermal and sound energy released into the environment.

How effectively a device, such as the car, transfers and transforms the total energy into useful energy is called **efficiency.** When people talk about how many miles per gallon their cars get, they are talking about efficiency.

You can modify the efficiency of an entire system by changing its individual components. For instance, tires alone can increase or decrease the efficiency of the car system by changing the amount of energy transferred between the tires and the road.

No energy transfer or transformation is 100 percent efficient. Although you can improve the design and operation of a device or action, there will always be energy transferred to the surroundings, or transformed to a form we are not going to use.

Friction

There are many forces in nature that will decrease the efficiency of a device, and one of those forces is friction. Friction refers to any force that resists the relative motion of an object.

Friction is not always a bad thing. The friction between a car's tires and the road allows the tires to grip the road and provides the means for the car to accelerate. Another use of friction occurs when you apply the brakes on a bicycle. Applying the brakes increases the friction on the wheels. This increase in friction causes the bike to quickly transform its mechanical energy into thermal energy, resulting in a decrease in speed of the bicycle.

Example

Cars are generally not very **efficient**. Gasoline contains a significant amount of potential energy. When it is burned within a car's engine, this stored energy is transformed into thermal energy. This released thermal energy expands air, which moves a piston, which turns a crank, which eventually is responsible for the spin of the car's wheels.

During each step of this process, some energy is transformed into unwanted thermal energy. A bit is even transformed into the sound of a revving engine! In many car engines, only about 15 percent of the energy stored in gasoline is transformed into the movement of the vehicle. The remaining 85 percent is transformed mostly into thermal energy that does not contribute to the movement of the car.

Math Connection

Efficiency is normally calculated as a percentage.

Efficiency = useful energy transferred / total energy used × 100

For example, a 60-watt incandescent light bulb uses 60 watts of total energy and gives off 9 watts of useful light energy.

Efficiency = 9 watts / 60 watts × 100 = 15%

A compact fluorescent light bulb only uses 11 watts of total energy to produce 9 watts of useful energy, making it more than 5 times more efficient.

Efficiency = 9 watts / 11 watts × 100 = 82%

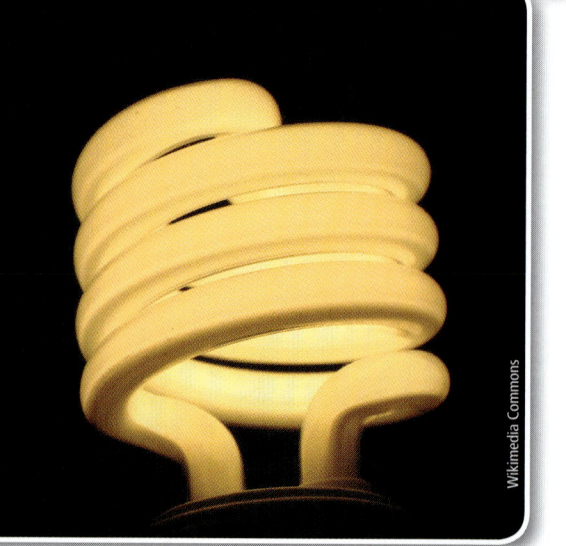

Compact fluorescent lamps (CFLs) and light-emitting diodes (LEDs) are more efficient options that transform a greater percentage of electrical energy into electromagnetic energy. Therefore, these lamps remain much cooler as they generate an equivalent amount of light.

Electrical Efficiency

Energy losses also accompany the flow of electrical current. You can sometimes detect this unwanted transformation to thermal energy by touching the side of your television set after it has been on for a few minutes; it may feel warm to the touch. At higher voltages, however, less electrical energy is transformed into thermal energy. That is why power transmission lines that distribute huge amounts of electricity are often energized to over 100,000 volts!

A lit incandescent bulb is extremely hot. The electricity that energizes the bulb is not transformed only into light. When objects such as the bulb's filament heat up, they emit electromagnetic energy. Unwanted thermal energy makes this a very inefficient use of electrical energy. These days, however, there are alternative types of lighting devices.

Fast Fact
We are really lucky that the energy transfers and transformations from the ocean bottom that produce tsunamis are very inefficient. Less than 10 percent of the energy from moving tectonic plates is actually transferred to the water above. Most of the energy is transferred and transformed into Earth's crust and interior as thermal energy and earthquake waves, sometimes affecting locations hundreds of miles away.

DART® System
Short for Deep-ocean Assessment and Reporting of Tsunamis, the DART® Project is part of an effort to save lives and property through monitoring undersea pressure changes generated by a passing tsunami. Each DART® system is deployed in a region that, in the past, has been an active tsunami site. The system is composed of two separate units: the tsunameter and its companion surface buoy. The tsunameter is located on the sea bottom and detects pressure changes associated with passing surface tsunamis. This data is then sent through sound signals to the moored surface buoy. The surface buoy relays this information to satellites that broadcast the data to monitoring stations on the ground.

DART® II Surface Buoy

Diameter: 2.5 m (8.2 ft)

Height (without superstructure): 1.8 m (5.9 ft)

Displacement: 4,000 kg (8,818 lbs)

Mooring Tether: 1.0–6.0 km (0.60–3.7 mi)

Anchor: 3,100 kg (6,834 lb)

Number of DART® systems deployed by the US: 39 buoys

Location of US DART® Systems: 31 buoys deployed in the Pacific Ocean (mostly along the Ring of Fire), 8 buoys deployed in the Atlantic Ocean, Gulf of Mexico, and Caribbean Sea.

Satellite Communication: Communicates with Iridium satellites using bi-directional transmission and control. It also receives and processes Global Positioning System (GPS) data.

Tsunameter Communication: Communication with seafloor component uses bi-directional sound signals.

DART® II Tsunameter

Detector Type: A bottom pressure recorder that collects temperature and pressure data

Sampling Interval (internal record): 15 seconds

Sampling Interval (event reports): 15 and 60 seconds

Sampling Interval (tidal reports): 15 minutes

Tsunami Data Report Trigger: Automatic trigger when the onboard data analysis suggests a tsunami event, or on-demand by warning center request

Sensitivity: At a depth of 6 km, the bottom pressure recorder can detect a change in bottom pressure generated by a tsunami with an amplitude of as little as 1 cm

Estimated Battery Life: Greater than 4 years

Mission 2: Waves of Change—Calculating Transfers and Transformations

Field Assignment

Exploring Energy Transfers and Transformations

Recall that your mission is to *understand how energy is transferred and transformed in predictable ways*. Now that you have been fully briefed, it is time to put your knowledge into action.

Dr. Titov knows how important it is for scientists to design and test their warning systems before putting them into use. His goal is to develop an operational tool that can provide quick and accurate tsunami forecasts in real-time. However, fast real-time forecast modeling of tsunami waves is complicated. First, he receives data about the location and magnitude of the seismic event. Then, he begins receiving data about the transfer and transformation of the earthquake energy by the tsunami wave from the network of DART® buoys in the ocean. After importing this data into his computer, a simulation, or model, is created which helps him make specific tsunami forecasts for locations around the world.

To begin this activity, you will analyze forecast data from Dr. Titov's simulation system that uses data from an event that occurred in the eastern Philippines. From this data set, you will make specific forecasts about the energy transfers and transformations for locations found within the Pacific Ocean.

After analyzing and forecasting the energy transfers and transformations in Dr. Titov's system, you will be faced with designing and testing your own energy transfer and transformation system. Using your knowledge of energy and how it can be changed, your challenge will involve designing, building, modifying, and testing a structure that will protect an egg when dropped from increasing heights.

Mission 2 Argonaut Field Assignment Video Join the National Argonauts as they study tsunamis with Dr. Vasily Titov.

Caution! Follow your instructor's safety protocols for dropping your egg protection structure.

Objectives: To complete your mission, accomplish the following objectives:
- Analyze tsunami propagation simulation data in the Pacific Ocean.
- Apply mathematical equations to develop a tsunami forecast for different locations.
- Apply mathematical equations to calculate potential energy and kinetic energy in order to help you design, build, and test a structure that will protect an egg when dropped from increasing heights.
- Evaluate the success of your structure and recommend ways to improve your design.

Materials
- Mission 2 Field Assignment Data Sheet
- graph paper
- raw eggs
- plastic bags
- math tools (p.114)
- tape measure
- balance
- stopwatch
- materials provided by instructor

Field Preparation

1. Examine the tsunami propagation map generated using tsunami simulation data from Dr. Titov.

2. Using another map source, such as an atlas or Google Earth™, determine the approximate distance from the Source Region to the following locations (in kilometers):

   a. San Francisco, California
   b. Point labeled Hawaii Region
   c. Point labeled South America Region
   d. Northern most point of New Guinea
   e. Tokyo, Japan

3. Using the propagation map, determine the amount of time it takes for the tsunami generated by the earthquake to reach the shores of each location.

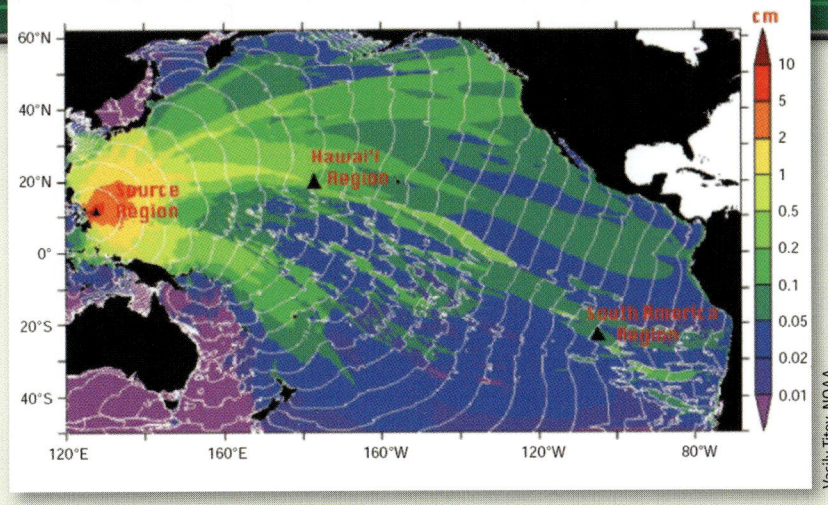

NOTE: White lines extending away from the source region represent 1 hour of elapsed time and wave heights are represented by color. (A larger version is available in the JMC.)

④ Using the following formula, determine the average speed of the energy propagated by the tsunami wave to each location from the source.

$$\text{average speed} = \frac{\text{total distance}}{\text{total time}}$$

⑤ From the propagation map, determine and record the approximate resultant wave height at each location.

⑥ Using the data collected, determine which areas, if any, need to be warned about a possible tsunami threat.

Mission Challenge

Your challenge will involve designing, building, modifying, and testing a structure that protects an egg from breaking as it is dropped from various heights. This will allow you to explore first hand the key variables involved with determining potential energy (PE), kinetic energy (KE), and protective system parameters.

① Using a sheet of graph paper, create a graph that shows the relationship between height and velocity for a dropped object using PE and KE and the Conservation of Energy. Graph height with a range from zero to five meters in 25 cm increments.

hint: PE = mgh;

KE = ½mv²;

PE = KE; solve for v

② Build and determine the mass of your egg protection system (with egg) using the materials provided by your instructor.

③ Calculate the PE of your protective system (with egg) at 50 cm increments from zero to five meters. Graph your data.

④ Test your device at 25 cm increments until your egg breaks.

⑤ Using the graph you created in step 1, estimate the velocity at which your egg broke. Using the graph created in step 3, determine the amount of PE it took to break your egg.

Mission Debrief

① Based on your results, beyond what velocity would you warn users of your egg protection structure not to exceed?

② How does this experiment relate to Vasily's research?

③ Compare your design with the designs of others in your classroom. Identify designs that performed well, and explain why you think they worked well. Consider designs that did not perform well, and suggest ways in which they could be improved.

Journal Question Explain how researchers like Dr. Titov use mathematical models to help save lives and protect property.

Connections

Math in Sports

Coefficient OF RESTITUTION

"It's a long fly ball, way back there! Watch it, watch it! It's out of here! That's right, another homerun for this year's champs!"

Although a game of baseball seems like a long way from a science classroom, it is not. The players might not be aware of it, but their actions, moves, and plays are all governed by a set of physical science relationships. Even their equipment is manufactured to strict standards that have been developed based upon the physical characteristics of balls and bats.

Every ball that is used in Major League Baseball must adhere to the same manufacturing and performance characteristics. Each ball has a cork core surrounded by layers of rubber. Woolen yarn is then wound tightly around this solid center. Covered by leather and sewn together with 108 stitches, the ball has a circumference between 22.86 cm and 23.50 cm (9.000–9.250 in.). It has a uniform mass between 142.0 g and 149.0 g (5.000–5.250 oz.).

The physical dimensions of size and mass are not the only standards to which the ball must conform. It must also have a coefficient of restitution (COR) that fits within allowed specifications. The COR is a mathematical measure of elasticity, or how well the ball will rebound when struck by a bat.

The COR can be determined using a basic calculation that compares the difference in velocities of the ball/bat system before and after the collision. COR is often abbreviated as *e*, and can be calculated using the following equation:

$$e = \frac{\text{post-impact velocity}_{\text{ball}} - \text{post-impact velocity}_{\text{bat}}}{\text{pre-impact velocity}_{\text{bat}} - \text{pre-impact velocity}_{\text{ball}}}$$

The equation may seem overwhelming at first, but it can be interpreted by looking at the range of possible values. What would happen if a ball made of mud struck a wall? Without any sort of bounce back, the ball would have zero post-impact velocity. The COR for such a collision would be 0.

The surface of a baseball is not the only determining factor of its coefficient of restitution. The internal contents and construction of the ball also play a significant part.

46 • Operation: Infinite Potential   www.jason.org

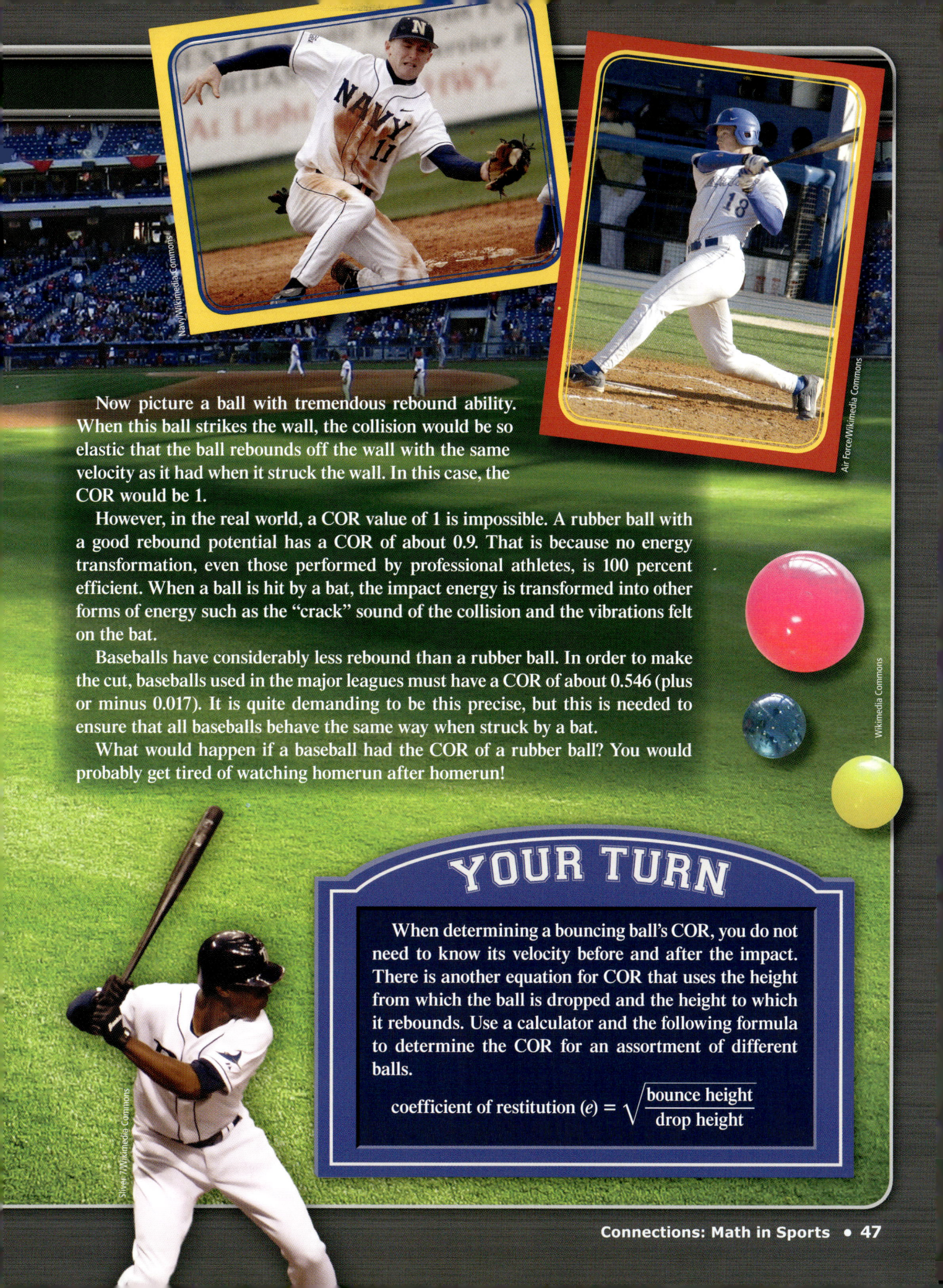

Now picture a ball with tremendous rebound ability. When this ball strikes the wall, the collision would be so elastic that the ball rebounds off the wall with the same velocity as it had when it struck the wall. In this case, the COR would be 1.

However, in the real world, a COR value of 1 is impossible. A rubber ball with a good rebound potential has a COR of about 0.9. That is because no energy transformation, even those performed by professional athletes, is 100 percent efficient. When a ball is hit by a bat, the impact energy is transformed into other forms of energy such as the "crack" sound of the collision and the vibrations felt on the bat.

Baseballs have considerably less rebound than a rubber ball. In order to make the cut, baseballs used in the major leagues must have a COR of about 0.546 (plus or minus 0.017). It is quite demanding to be this precise, but this is needed to ensure that all baseballs behave the same way when struck by a bat.

What would happen if a baseball had the COR of a rubber ball? You would probably get tired of watching homerun after homerun!

YOUR TURN

When determining a bouncing ball's COR, you do not need to know its velocity before and after the impact. There is another equation for COR that uses the height from which the ball is dropped and the height to which it rebounds. Use a calculator and the following formula to determine the COR for an assortment of different balls.

$$\text{coefficient of restitution } (e) = \sqrt{\frac{\text{bounce height}}{\text{drop height}}}$$

Connections: Math in Sports • 47

Mission 3

Power to the People
The Current State of the Grid

"We are using a lot of fossil fuels, and our real mission here is to come up with ways to do that safely, to do that cleanly, and to not impact people's environment."

—Dr. Larry Shadle
Research Group Leader,
Model Validation Research Group,
National Energy Technology Laboratory

Larry Shadle

Larry Shadle and his research team use advanced laser Doppler technology to analyze mixtures of coal emissions and sorbents used in coal-fired power plants. Through this technology, he hopes to develop a system to minimize the toxic compounds and greenhouse gases that are released into the environment every day.

Meet the Researchers Video
Discover what motivates Larry and how his work is reducing the environmental impacts of coal-fired power plants.

Model Validation Research Group Leader, NETL

Read more about Larry online in the JASON Mission Center.

Photos above (left to right): Marcus Obal/Wikimedia Commons; Michael Wolf/Wikimedia Commons; Marya/ Wikimedia Commons; Wikimedia Commons; Peter Haydock, The JASON Project; Guido Gerding/Wikimedia Commons

Your Mission...

Investigate how we currently meet our energy and power needs.

To accomplish this mission successfully, you will need to

- Survey the history of energy.
- Discover the relationship between electricity and magnetism.
- Investigate how electrical energy travels in circuits.
- Explore the generation and distribution of electricity.
- Understand how nuclear and hydroelectric resources are used to generate power.
- Analyze the role of fossil fuels in energy and power systems.

Mystery Connection

ALERT!

Argonauts, the world is in crisis, and we need your help! Here is the situation. Based on our government research:

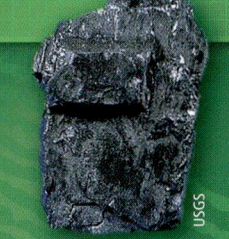

- **We may have only 150 years of coal resources left.**
- **Projections from the United States Department of Energy show that demand for coal is likely to increase 67 percent over the next 20 years.**
- **Accordingly, carbon dioxide emissions are projected to increase by the same percentage over the next 20 years. Carbon dioxide has been linked to global warming and other environmental problems.**
- **Researchers are working on alternative forms of energy, but it is uncertain how long it will take to develop these energy forms and change our world's infrastructure to take advantage of the new technology!**

How can we use all the resources at our disposal to meet our current energy needs in an environmentally responsible manner?

Mission 3: Power to the People—The Current State of the Grid • 49

The Current State of the Grid

Entering the building, Larry Shadle puts on his hard hat, adjusts his safety glasses, and grabs a wrench as large as his arm. He bends down next to a metal pipe and gives it a whack with the wrench. A sandy sludge oozes out of the pipe and lands in a bucket below. Shadle looks up at the massive simulator which his team uses to learn how to better capture the carbon dioxide (CO_2) that coal-fired power plants emit.

The mixture races through clear plastic tubes, climbing five stories to the very top of the building before being sent through a giant cyclone and then back to ground level. On one floor, technicians focus a laser beam on the mixture to understand the behavior of the particles as they rise and fall. On another floor, a scientist photographs the mixture with a high-speed video camera so sophisticated that you can actually see the patterns of the particles as they move through the tubes. Everywhere, the steady hum of machinery is heard.

For 20 years, Larry Shadle has dedicated himself to this lab, located at the National Energy Technology Laboratory (NETL) in Morgantown, West Virginia. The lab is an enormous simulation of cutting-edge technology that he hopes to implement in coal-fired power plants all over the world. Here, Larry and his colleagues develop carbon-control technology that will allow people to utilize coal resources far more cleanly and efficiently than ever thought possible. While some scientists work on other forms of energy, such as solar, wind, and nuclear, Larry is figuring out how we can use our available coal resources while preserving our planet's environment.

Mission 3 Briefing Video Prepare for your mission by viewing this briefing on your objectives. Learn how scientists like Larry Shadle search for ways to reduce the impact fossil fuels have on the environment.

Mission Briefing

The History of Energy

What is an **energy portfolio**? Throughout history, people have generated and used energy to live. This energy often comes from multiple sources that may include wind, water, fossil fuels, or the sun. These energy options, and how we choose to use them, make up a person's or a region's energy portfolio. Over time, the world's portfolio changes based upon the availability of resources and our technological ability to use them.

Think of all the ways in which you use energy in your daily life to make things work for you. Just today, chemical energy from gasoline may have powered a car or bus to take you to school. You probably used electrical energy to turn on a light or charge a cell phone. Later, thermal energy may be used to cook your dinner. In almost everything we do, we rely upon energy. Can you imagine what life would be like if you did not have gasoline to get around or electricity to power a light?

There was a time when people had fewer energy options than we have today. In fact, several hundred years ago, people would burn wood and other forms of plant material in order to meet most of their daily energy needs.

The Discovery of Fuel

People have always obtained energy from their food, but fire gave them the ability to harness energy in new ways. Evidence suggests that as many as 1.5 million years ago, early humans may have used fire to cook food, keep warm, and frighten animals away.

These fires used wood and plant materials for fuel. A **fuel** is a substance that is consumed to produce energy. Remember that plants transform electromagnetic energy from the sun into chemical energy that is stored within plant tissue. When plant material is burned, the molecules react with oxygen. During this chemical change, bonds are rearranged and potential energy is released in the forms of heat (thermal energy) and light (electromagnetic energy). This process of burning a substance to release stored energy is called **combustion.**

Sun (electromagnetic energy) ➡ Plants (chemical energy) ➡ Fire (thermal energy and electromagnetic energy)

The discovery and use of fossil fuels set the stage for the industrialization of the world. **Fossil fuels** are carbon-based fuels, such as coal, oil, or natural gas, that are formed by the compression of plant and animal remains.

Coal burns hotter and longer than wood and soon became a preferred energy source. People started using coal regularly to heat their homes. During the colonial days, farmers in England and other parts of Europe would dig coal up from outcrops exposed at Earth's surface.

Join the Team

At the National Energy Technology Laboratory in Morgantown, West Virginia, Dr. Larry Shadle shows Argonauts (L to R) Jackie Martin, Tim West, Hiyam Añorve Garza, and Melissa Hall how new technology can reduce carbon emissions from the burning of coal. Through his research, Dr. Shadle hopes to implement carbon-control technology that will allow us to use the coal resources available, while minimizing the impact on the environment.

Fast Fact

Humans have always been innovative, discovering different ways of harnessing energy from nature. For instance, over 5,000 years ago, ancient Sumerians and Egyptians exploited the power of wind with sailboats. Evidence shows that horses were domesticated up to 5,600 years ago, tapping into their energy for travel and work. The first clear evidence of a water wheel points to the ancient Greeks, who used it to grind grain over 2,000 years ago. All of these energy sources continue to be used in some form today. However, as technology has evolved, new and more powerful forms have emerged. Will you be able to help in the discovery of the next major energy innovation?

In the early 1700s, however, England faced an energy crisis. Most of the surface coal had been consumed. Geologists knew that much more coal lay far beneath the surface, but it was dangerous and difficult to mine. Mines dug deep in the earth would easily fill up with water, putting miners in danger and making extraction of coal difficult. People had to come up with something new in order to obtain the energy they needed.

The Steam Engine

In 1769, a Scottish inventor named James Watt patented a steam engine that pumped the water out of mines and enabled the large-scale mining of coal.

How did Watt's steam engine work? First, coal was used to boil water. The resulting steam expanded and exerted pressure to move a piston upward.

Once that piston reached a certain height, a valve opened and cold water was released to cool the steam. Cooling the steam caused it to condense, take up less space, and therefore released the pressure it was exerting on the piston. This caused the piston to come back down again. Alternately creating and cooling steam produced an up-and-down piston motion—the mechanical energy of the early steam engine. Connecting the piston to a pump allowed water to be removed from mines, where far more coal could be obtained than what was needed to power the engine.

Coal (chemical energy) ➡ Boiling water (thermal energy) ➡ Steam (mechanical energy) ➡ Pump (mechanical energy)

With access to a huge amount of energy-rich coal from deep within the earth and with a powerful new steam engine, an explosion of technology followed.

Once Watt and others could use steam pressure to move a piston, they were soon able to add other mechanical devices to enable new forms of transportation. For instance, crankshafts converted the piston's up-and-down motion into a rotational motion that could turn a wheel. This technology was used to power steam engines and steam boats.

The Internal Combustion Engine

The discovery of another fossil fuel—oil—enabled other, more advanced modes of transportation. Oil is a form of **petroleum**, or "rock oil," which includes natural gas and tar. Like coal, petroleum is a fossil fuel, but it is formed by the compression of plants and animals that once lived in a marine environment.

As early as 3000 B.C., early Mesopotamians used petroleum to make things, such as medicine and roads. In the mid-1800s, oil began to be used more and more for heat and light in the United States. But the U.S. petroleum industry really began in 1859, when a Pennsylvania railroad conductor named "Colonel" Edwin L. Drake struck oil with a homemade rig that drilled down 21.3 m (70.0 ft) into the earth.

The ability to refine petroleum greatly enhanced the development of modern automobiles. When crude oil is refined, petroleum products like gasoline, diesel, and propane are separated for use. Though the earliest automobiles were steam powered, the development of gasoline-powered cars began by the end of the 1800s. By 1908, Henry Ford's assembly lines were mass-producing automobiles that relied on gasoline.

Fast Fact

The first documented steam engine was developed in the first century A.D. by a Greek mathematician named Hero of Alexandria. Hero's "aeolipile" consisted of an air-tight chamber that could rotate on a bearing, with curved nozzles sticking out of its sides. Water was heated either in the chamber or in a separate basin to produce steam that was piped into the chamber. The force of the steam through the pipes caused the chamber to spin around.

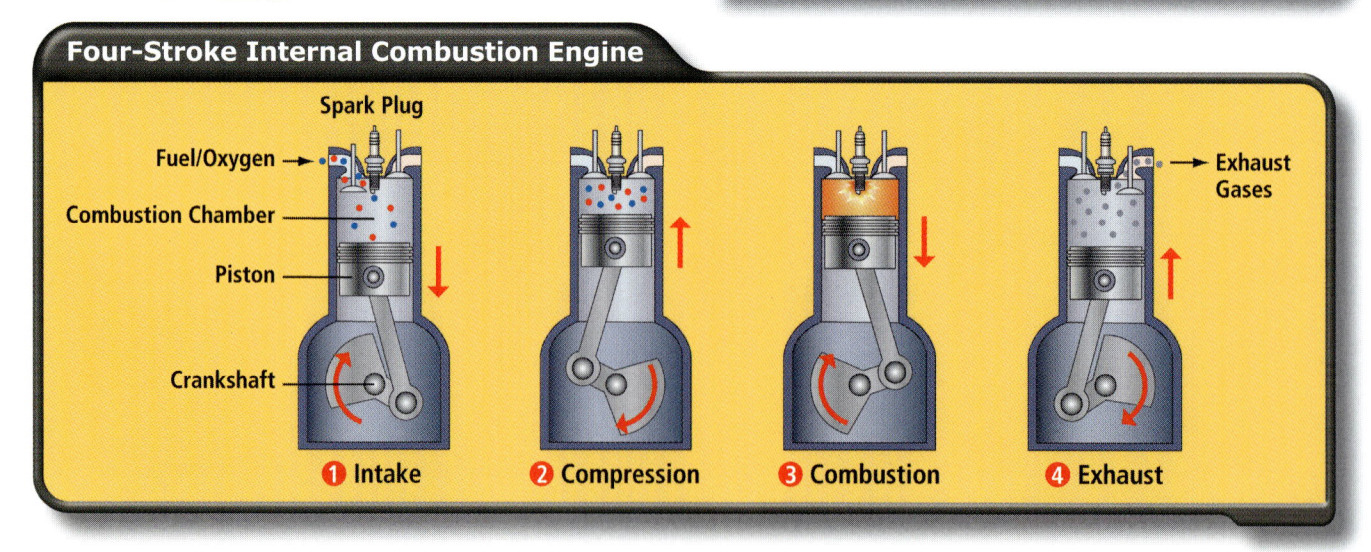

Four-Stroke Internal Combustion Engine

Spark Plug
Fuel/Oxygen
Combustion Chamber
Piston
Crankshaft
Exhaust Gases

❶ Intake   ❷ Compression   ❸ Combustion   ❹ Exhaust

Operation: Infinite Potential   www.jason.org

Typical of many early cars, today's automobiles also use an **internal combustion engine.** In many ways, the internal combustion engine and Watt's steam engine are similar. In each case, the burning of a fuel causes a gas to expand and move a piston. However, in the case of the internal combustion engine, the fuel is gasoline, which is combusted inside the chamber, causing the pistons to move.

> **Gasoline** (chemical energy) ▶
> **Combustion** (thermal energy) ▶
> **Expansion of gas** (mechanical energy) ▶
> **Movement of piston** (mechanical energy)

The Turbine Engine

In addition to engines that use pistons, inventors found a way to produce engines that used turbines. A **turbine** engine consists of a series of blades arranged in a circle, like a wheel. A fluid, such as water, steam, air, or a combustible gas, pushes against these blades, going in one side and coming out the other. In this way, the wheel is made to turn around. A windmill is an example of a turbine.

At the end of the 19th century, many types of steam turbines began to emerge for a variety of purposes. As technology advanced, turbines were adapted to utilize various energy sources. Today, gas-powered turbines are used to power jet airplanes. Other turbines are used in various ways, such as in manufacturing, on helicopters, and even in space shuttles. Steam-powered turbines are still one of the main ways we generate electricity today.

Fast Fact

In the early 1800s, engineers proposed a design that directed steam straight onto the blades of a turbine. Unfortunately, steel was not yet strong enough to hold up to the stress of such rapid rotation. In 1884, a British engineer named Charles Algernon Parsons used new steel technology that allowed a turbine to spin significantly faster than ever before. As a result, in the early 1890s his steam turbine was used in an actual power station to generate electricity.

Laser Doppler Velocimeter

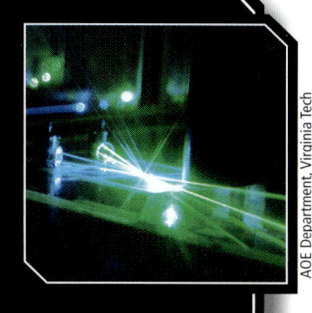

The primary objective of emissions-controlled technology is to minimize the toxic compounds and greenhouse gases (CO_2, NO_x, SO_2, CO, and CH_4) produced by coal-fired power plants.

A technological advance that addresses this issue is the fluidized bed. A fluidized bed creates a floating mix of finely crushed coal, ash, and sorbent particles. Harmful gases produced during the combustion process react both physically and chemically with sorbent particles and are captured and removed. In order to remove the toxic gases with maximum efficiency, it is critical that the components of the fluidized bed be accurately monitored.

Dr. Larry Shadle and his team of scientists at the National Energy Technology Laboratories (NETL) use a Laser Doppler Velocimeter (LDV). This machine applies advanced laser Doppler technology to help monitor flow structure and mixing characteristics in a fluidized bed. It is important to understand this mixing because it helps to optimize designs for future fluidized beds.

Using information from the LDV, Larry and his team are developing detailed computer models that describe the physics of the mixing process. These models can then be applied to any application that uses fluidized beds.

Laser specifications: Ion Laser Technology Model 5500A

Laser power supply: Single Phase 220 V 20 A

Laser beam properties: Wavelengths (green): 514.5 nm; Wavelengths (blue): 488.0 nm; Focal length, L: 750 nm; Beam separation: 40 nm; Laser beam diameter: 2 nm

The Modern Energy Portfolio

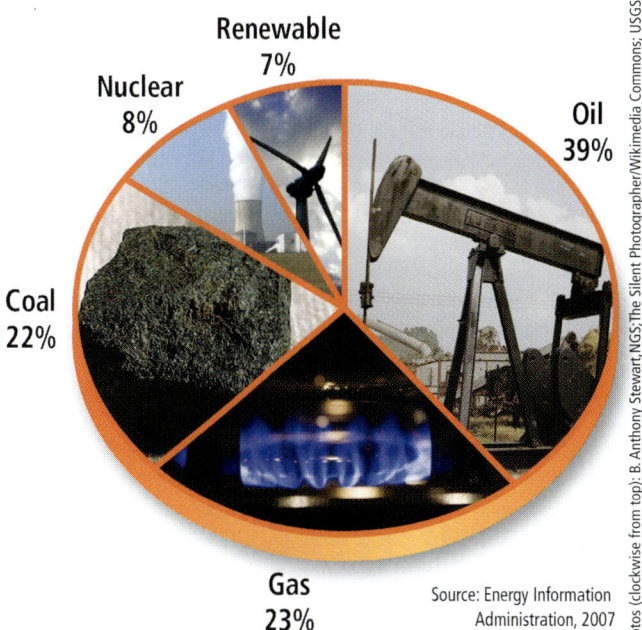

- Renewable 7%
- Nuclear 8%
- Oil 39%
- Gas 23%
- Coal 22%

Source: Energy Information Administration, 2007

Mission 3: Power to the People—The Current State of the Grid • 53

Generating Electricity: Magnets

As engine and turbine technology advanced throughout the end of the 19th century, machines which could produce electrical energy, called **generators**, were starting to be developed. The development of these machines, and the understanding of how electrical energy can be generated, transformed the way we have lived our lives ever since.

To produce electrical energy we can use every day, we must first understand the underlying components of electricity generation: magnets and circuits.

Chances are, you've seen magnets in action before; but do you know how they work? To understand magnetism, you need to take a closer look at atoms. An atom is the smallest unit of an element having all of the characteristics of that element. Atoms consist of protons and neutrons together in a core nucleus and electrons moving outside the nucleus. Protons have a positive charge, neutrons have no charge, and electrons have a negative charge.

The movement of any charged particle, including an electron, creates a **magnetic field**. A magnetic field is an area surrounding an object where magnetic forces are exerted.

Magnetic Field Lines Around a Magnet

Fast Fact

Atoms have an equal number of protons and electrons. The positive and negative charges cancel each other out. However, when the number of protons and electrons are unequal, the atom assumes a net, or overall, charge.

Electrostatic charges are observed as pushes and pulls between these charged objects. If two similarly charged objects are brought together, such as two negatively charged balloons, they will push apart.

Charges can be transferred when two dissimilar materials are brought in close contact with each other. For instance, when fur is rubbed against a balloon, the electrons will move from the fur to the balloon, causing the objects to be attracted to one another.

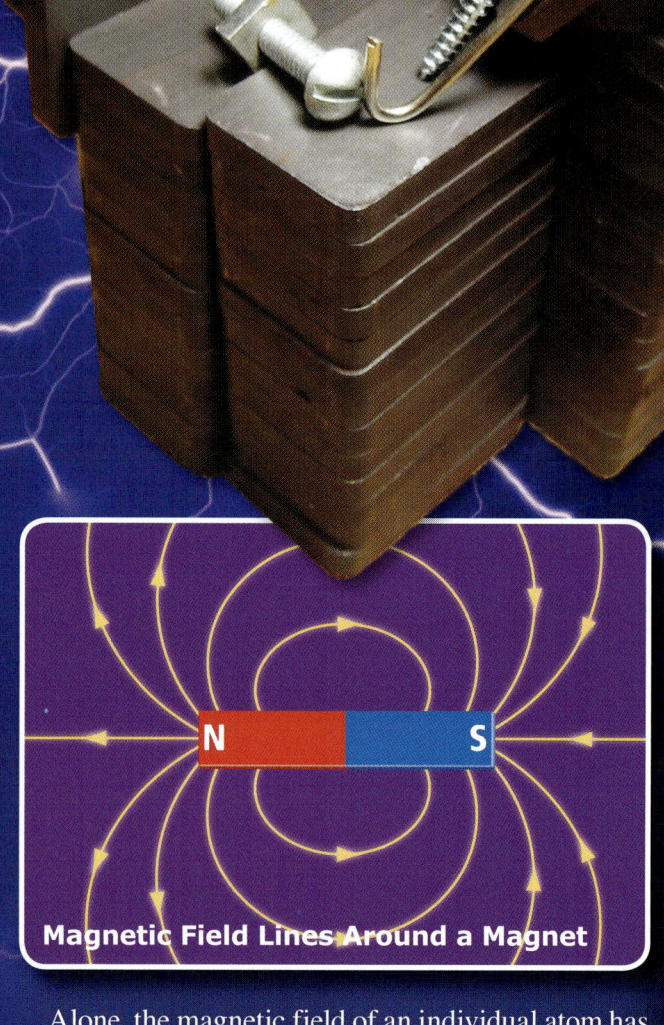

Alone, the magnetic field of an individual atom has a negligible effect on its surroundings. However, when fields of neighboring atoms align, the effect on its surroundings is increased. When many fields align, as in a magnet, an observable magnetic field is produced.

Every magnet has two poles—a north pole and a south pole. A magnetic force flows from the north to the south pole. This force is what produces the observed magnetic field. The poles are the strongest part of the magnet, where the magnetic field is most concentrated. This increased force is responsible for the observed repulsion, or push apart, between like poles and attraction between unlike poles.

Lab 1

Exploring Magnetism

Larry Shadle is well aware of coal's legacy in electricity generation. Although coal was burned as a heat source for thousands of years, it was not until 1882 that this fuel was used to meet the demand for electricity. At that time, another energy scientist, Thomas Edison, presented the first centralized electricity station. When powered up, the station's generator (then called a dynamo) created enough electricity to power 1,200 light bulbs! Edison's choice of station fuel was coal—the same fossil fuel that meets over half of the current electrical energy demands of the United States.

Like today's power plants, Edison's electricity station was an application of the relationship between magnetism and electricity. What are these energy forms, and how do they interact? In this activity, you will explore magnetism and electricity. Then you will assemble an **electromagnet** and analyze its operation.

Materials
- Lab 1 Data Sheet
- 2 bar magnets
- steel paper clips
- connecting wire with alligator clips
- sheet of heavy stock paper
- iron filings
- iron nails
- magnetic compass
- D-cell in battery holder
- horseshoe magnets, disk magnets or other magnets provided by instructor

⚠️ When using iron filings, make sure to always wear safety goggles and protective gloves. Also make sure that all fans are turned off and windows are shut.

Lab Prep

1. Place two magnets end-to-end. How do the magnets behave? Explain your observations in terms of *like* and *unlike* poles. Now turn one magnet 180 degrees so that its other end faces the first magnet. What happens? Are the observed forces larger or smaller at the poles? How can you tell?

2. Next, use one of the magnets to attract a steel paper clip. Must the magnet be touching the clip in order to affect the clip's position and movement? Explain.

3. To magnetize a paper clip, attach it to the edge of a magnet. Using only the paper clip tip, pick up another paper clip. How many paper clips can you pick up with your magnetized clip? Compare your results with your classmates.

Make Observations

1. Put on your safety goggles and gloves. Lay a bar magnet flat on the desktop. Cover it with a sheet of heavy stock paper.

2. With the magnet positioned beneath the center of the paper, gently sprinkle iron filings on the paper's surface. Continue until a distinct pattern emerges. Draw this pattern of filings. How does this pattern relate to the invisible magnetic field?

3. Move the compass around the bar magnet's magnetic field. How does the compass needle align in the field? Based upon your observations, how can a compass be used to identify a magnet's poles? Explain.

4. Predict the field shape between two magnets that are placed with like poles separated by a small distance. How might the field change if one magnet was rotated so that the two opposite poles faced each other? Draw and explain the difference you see.

5. Predict the shapes of the magnetic fields of horseshoe magnets, disk magnets, or other magnets provided by your instructor. Use iron filings to test your predictions.

6. Make an electromagnet. Wrap a length of wire around an iron nail, forming a coil with at least ten turns. Test this nail for magnetic properties. Connect the free ends of the wire to the terminals of a D-cell battery. What happens now?

7. Use a compass to identify the poles of the electromagnet. Does the direction of the coil wrap affect the identity of the poles? Explain.

Journal Question From what you discovered about poles, what are the advantages and limitations of a horseshoe shaped magnet?

Simple Circuits

Series Circuit — Power Source, Load 1, Load 2

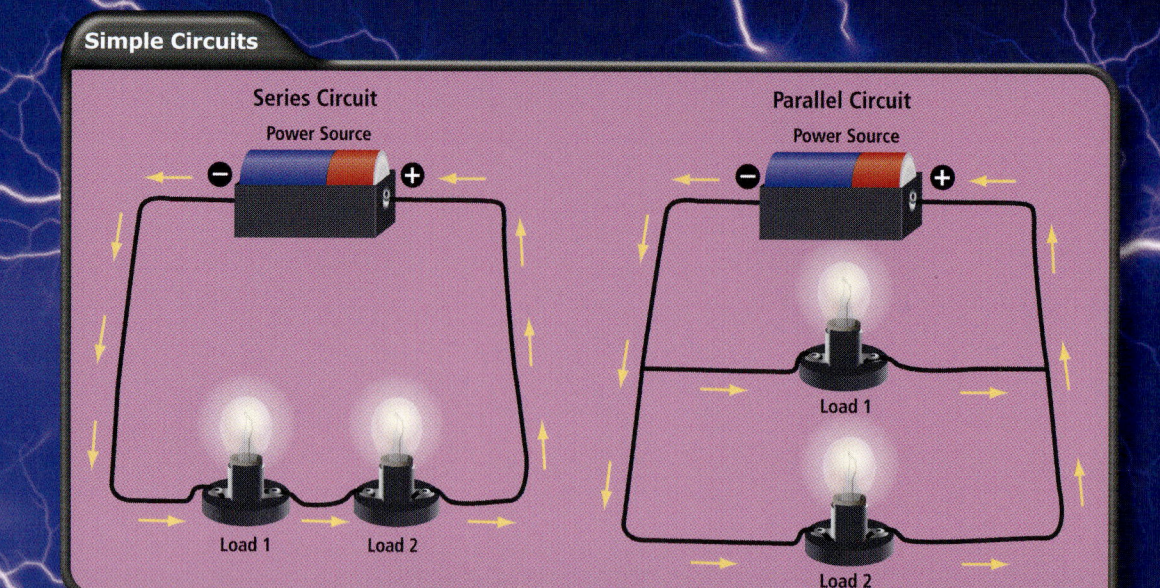

Parallel Circuit — Power Source, Load 1, Load 2

Generating Electricity: Circuits

Sometimes charges can be set in motion, flowing along a path. This flow of charges is called an **electric current**. Some materials allow charges to flow easily. They are called **electrical conductors**. Other materials that do not allow charges to flow easily are known as **electrical insulators**.

In order to get charges flowing, you need some sort of "pump" to push electrons. A household battery is an example of a device that can push electrons. However, to produce a continual flow of charge, you need a complete path, or **circuit.**

Battery (chemical energy) ➡ **Current** (electrical energy) ➡ **Light bulb** (electromagnetic and thermal energy)

Series Circuit

There are two basic types of circuits. Series circuits offer only one possible path for charges to flow. In a series circuit, all of the current flows through each of the circuit components. If there is a break in the path at any location, all of the charges stop moving.

Parallel Circuit

The second type of circuit is called a parallel circuit. Parallel circuits offer more than one route for the flow of charges. If the path in one branch becomes incomplete, the current still flows in the other circuit branches. This type of wiring can be seen in some holiday light sets in which one light may burn out, but the rest remain lit.

Fast Fact

What's in a name? The word *volt* is named after the Italian scientist Alessandro Volta. He is credited with building the world's first battery out of zinc, silver, and cardboard soaked with salt water.

As long as there is a "pump" and a closed circuit, electrons will continue to flow. In a battery, electrons will flow from the negative terminal toward the positive terminal, creating a direct current of electricity. **Direct current** (DC) electricity occurs when charges flow in only one direction through a circuit.

Ohm's Law

Voltage is the difference in electric potential between two locations. The measure of how easily charges flow through a substance or device is its **resistance**. Voltage, electric current, and resistance are very important when considering the design of circuits. **Ohm's Law** describes the relationship between voltage, electric current, and resistance. It states that as voltage goes up, current goes up, and as resistance goes up, current goes down.

Ohm's Law

$$V = IR$$

V = **voltage** (in volts)
I = **current** (in amperes)
R = **resistance** (in ohms)

Lab 2

Series and Parallel Circuits

Larry Shadle is researching how to reduce emissions from coal-fired power plants. Coal-fired power plants are one way of generating the electrical energy that flows through circuits in your home, powering devices for everyday use.

In this activity, you will construct a series circuit and a parallel circuit. You will then have the opportunity to explore these electrical paths using a multimeter. As you collect data, you will discover differences in these circuit types.

Materials
- Lab 2 Data Sheet
- connecting wire with alligator clips
- D-cell battery holder
- math tools (p.114)
- D-cell battery
- 2 1.5-volt lamps
- 2 lamp holders
- multimeter

Lab Prep

1. Review Ohm's Law in the Math Tools section. Summarize the relationship expressed by current, voltage, and resistance.

2. Insert the battery and lamps into their holders.

3. Review the operation of a multimeter. Set it to measure resistance, and place the leads on the two contacts of the lamp holder. Measure and record the resistance of this lamp. Using the same procedure, measure and record the resistance of the second lamp.

4. Predict the resistance of two lamps that are connected with a single wire. Then, test your prediction.

5. Set the multimeter to measure DC voltage, and place one lead on each of the two battery holder terminals. Measure and record the voltage generated by this cell.

Make Observations

1. Assemble a basic series circuit using one battery, one lamp, and connecting wires.

2. Set the multimeter to measure current. Undo one of the connections, and insert the multimeter into the circuit. Measure and record the current that flows through the circuit.

3. Is the current the same throughout the circuit? Test your hypothesis.

4. Using Ohm's Law, predict how adding a second lamp in series will affect the current and lamp brightness.

5. What happens to the current and lamp brightness when a second lamp is added? Explain.

6. Predict what will happen if you unscrew one of the lamps. Remove the lamp and see what happens. Explain your observations in terms of complete circuit paths.

7. Disassemble the series circuit. Reassemble these components into a basic parallel circuit using one battery, two lamps, and connecting wire.

8. Use your multimeter to measure the current between the cell and the first lamp. Then, measure the current between the first lamp and the second lamp. Are they the same value? If not, explain the difference.

9. Predict what will happen when one of the lamps is removed from the circuit. Remove the lamp and see what happens. Explain your observations in terms of complete circuit paths.

Extension

According to Ohm's Law, what would happen if you connected two batteries together in parallel? In series? Try it, and measure the effect on the circuits with the multimeter.

Journal Question What are some of the possible advantages and limitations of series and parallel circuits? Describe applications of these types of circuits within your home.

Mission 3: Power to the People—The Current State of the Grid • 57

Coal Power Plant

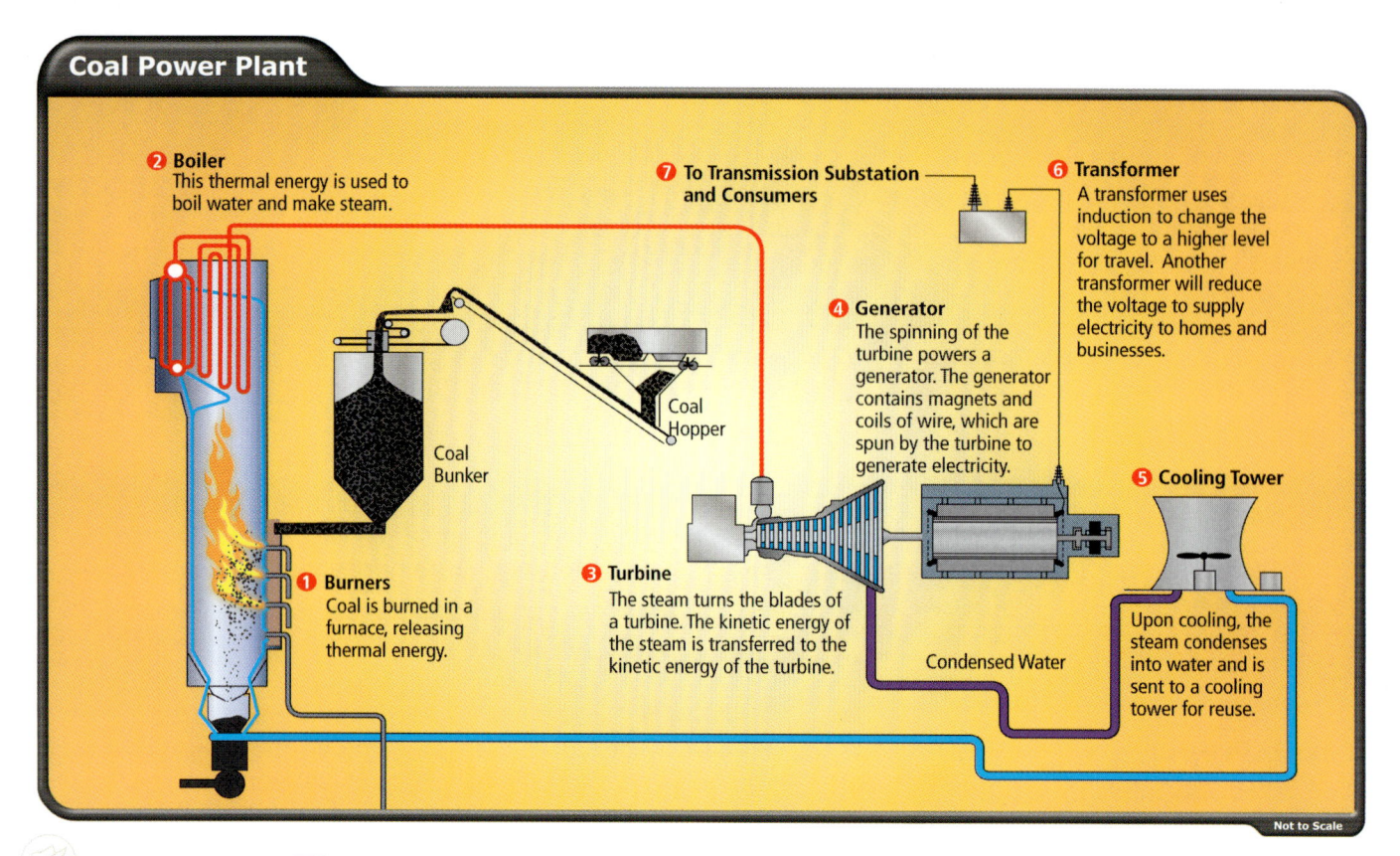

② Boiler
This thermal energy is used to boil water and make steam.

⑦ To Transmission Substation and Consumers

⑥ Transformer
A transformer uses induction to change the voltage to a higher level for travel. Another transformer will reduce the voltage to supply electricity to homes and businesses.

④ Generator
The spinning of the turbine powers a generator. The generator contains magnets and coils of wire, which are spun by the turbine to generate electricity.

Coal Hopper

Coal Bunker

⑤ Cooling Tower
Upon cooling, the steam condenses into water and is sent to a cooling tower for reuse.

① Burners
Coal is burned in a furnace, releasing thermal energy.

③ Turbine
The steam turns the blades of a turbine. The kinetic energy of the steam is transferred to the kinetic energy of the turbine.

Condensed Water

Not to Scale

Generating Electricity: Generators and Power Plants

In the early 1800s, Michael Faraday discovered that if a magnet is moved inside a coil of wire, electrical current flows through the wire. This happens when the magnet exerts a force that causes charges to move through the surrounding wire coil.

Based upon Faraday's discovery, today, people build generators to produce electrical energy for everyday needs. A **generator** is a device that transforms mechanical energy into electrical energy. Electrical current is induced in a wire that is exposed to a changing magnetic field. It does not matter whether the wire or magnet is in motion, as long as the magnetic field across the wire keeps changing.

All generators initially produce **alternating current** (AC), an electric current that reverses its direction within a circuit at regular cycles. The cycling occurs when the north and south orientation of the magnets changes with respect to the wire. When wired to devices that change AC to DC, a generator can produce DC current.

Power plants use generators to transform an initial energy source into electrical energy. Today, nuclear, hydroelectric, and fossil fuels like coal and gas generate the majority of the world's electrical energy. While individual power plants are fueled in a variety of ways, most use a primary energy source to spin a turbine that powers an electric generator.

Example

In power plants, we utilize the relationship between electrical energy and magnetism to generate electricity. We can also use electrical energy to induce magnetism in some metals, making an **electromagnet**. An electric current moving through a wire creates a magnetic field around the wire. This field can be concentrated by wrapping the wire into a tight coil. An iron object, such as a nail, inserted into the center of the coil can produce an even stronger electromagnet with an even greater magnetic effect.

Basic AC Generator

Magnet — S N
Coils
Magnet — S N
Brush
Slip-Rings
Brush

Not to Scale

58 • Operation: Infinite Potential www.jason.org

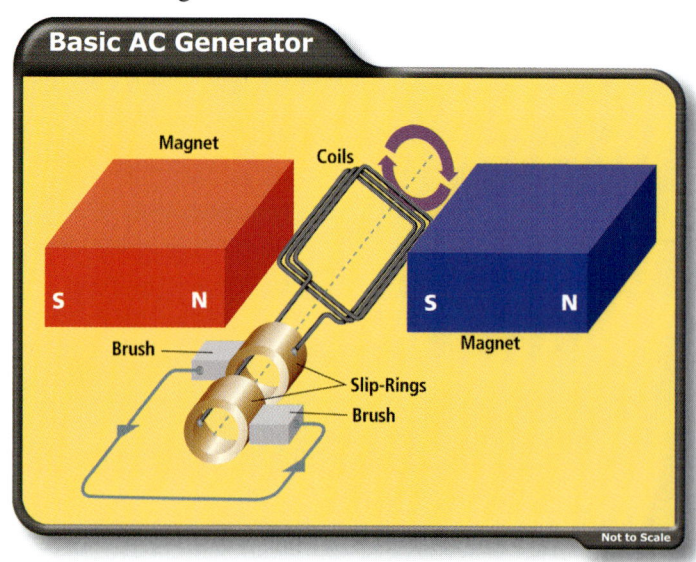

Lab 3

Generating Electricity

Although our country's coal reserves are substantial, we need to improve the way this fuel is transformed into electrical energy. Engineers and researchers, such as Larry Shadle, accept this challenge and are devising and testing innovations in coal technologies.

In this activity, you will explore the relationship between magnetism and electricity. Then, you will consider ways in which the efficiency of electricity-generating devices might be improved.

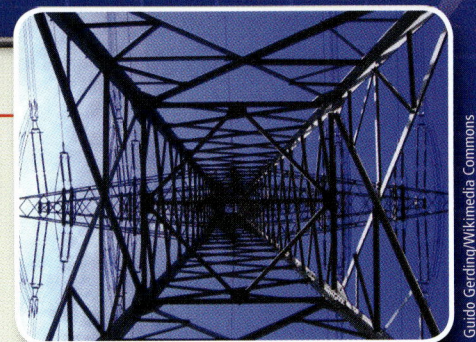

Materials
- Lab 3 Data Sheet
- 150 cm of 22 or 26 gauge magnet wire
- sandpaper
- multimeter
- cardboard tube
- strong rectangular magnet
- electricity generator tool (p.115)
- 2 connecting wires with alligator clips

Lab Prep

1. Obtain a length of coated wire about 150 cm in length. Sand the enamel coating from both ends of this wire.

2. Bend the wire into a large, flat loop.

3. Use connecting wires with alligator clips to attach the exposed ends of this wire to the leads of a multimeter.

4. Set the multimeter to measure DC voltage. Does the tool detect any voltage in the wire? Explain.

5. Hold a magnet still at a variety of locations both inside and outside the coil of wire. Record the voltage in each location.

6. Move the magnet in a variety of ways both inside and outside the coil of wire. Record the voltage each time you change location or motion.

7. Undo the connections, and wrap the wire around a cardboard tube forming a tight coil of wire. Leave the ends of the wire free, projecting out from the coil. Use a piece of masking tape to secure the coil shape.

8. Repeat steps 5 and 6, recording your observations.

9. Which combination of loop size, magnet location and motion-type produced the maximum voltage? Explain.

10. Based on what you have observed so far, compile a list of the factors that could affect the output or efficiency of electricity generation. Design and perform an experiment that investigates how changing one of these factors affects output or efficiency.

Make Observations

1. Create the electricity generator as indicated in the Tools section at the back of the book.

2. Adjust the assembly so that the magnets align within the wire coil. Spin the magnet/pencil assembly. Note the role that inertia plays in maintaining rotation. How is inertia an advantage? How is it a limitation?

3. Use alligator clips on the connecting wires to attach the sanded ends of the coil to the leads of the multimeter. Turn on the meter and set it to measure DC voltage.

4. What happens when you spin the inner assembly? How does spin speed and direction affect output?

5. Consider ways to improve the efficiency of your generator. Make a list of changes that might increase the output of electrical energy. Include a list of materials and a blueprint. With your instructor's permission, build your new and improved generator and test it.

Journal Question Compare and contrast a generator with an electrical motor.

Mission 3: Power to the People—The Current State of the Grid • 59

Power Distribution Grid

1 Power Plant
Electrical energy is generated at a power plant.

2 Transmission Substation
Transformers increase the voltage and send electrical energy out over transmission lines.

3 High-Voltage Transmission Lines
Typical voltages for long-distance transmission are anywhere from 155,000 to 765,000 volts. Increasing the voltage minimizes the loss of energy as the electric current flows over the line.

4 Power Substation
Transformers reduce the voltage to prepare electrical energy to be sent to users. Circuit breakers and switches allow a substation to disconnect itself from the transmission grid or from individual distribution lines.

5 Distribution Bus
Electrical energy is split out in multiple directions, so that it can be sent to homes over distribution lines.

6 Transformer
Transformers decrease the voltage of the electrical energy to 7,200 volts to travel along distribution lines.

7 Distribution Lines
Distribution lines can be above or below ground.

8 Transformer Drum
Transformers convert electricity to 240 volts, which enters your home through two 120-volt distribution lines.

9 Consumer
Some appliances in your house use 120 volts, such as your television, hair dryer, and blender. Others, such as your oven and clothes washer use 240 volts.

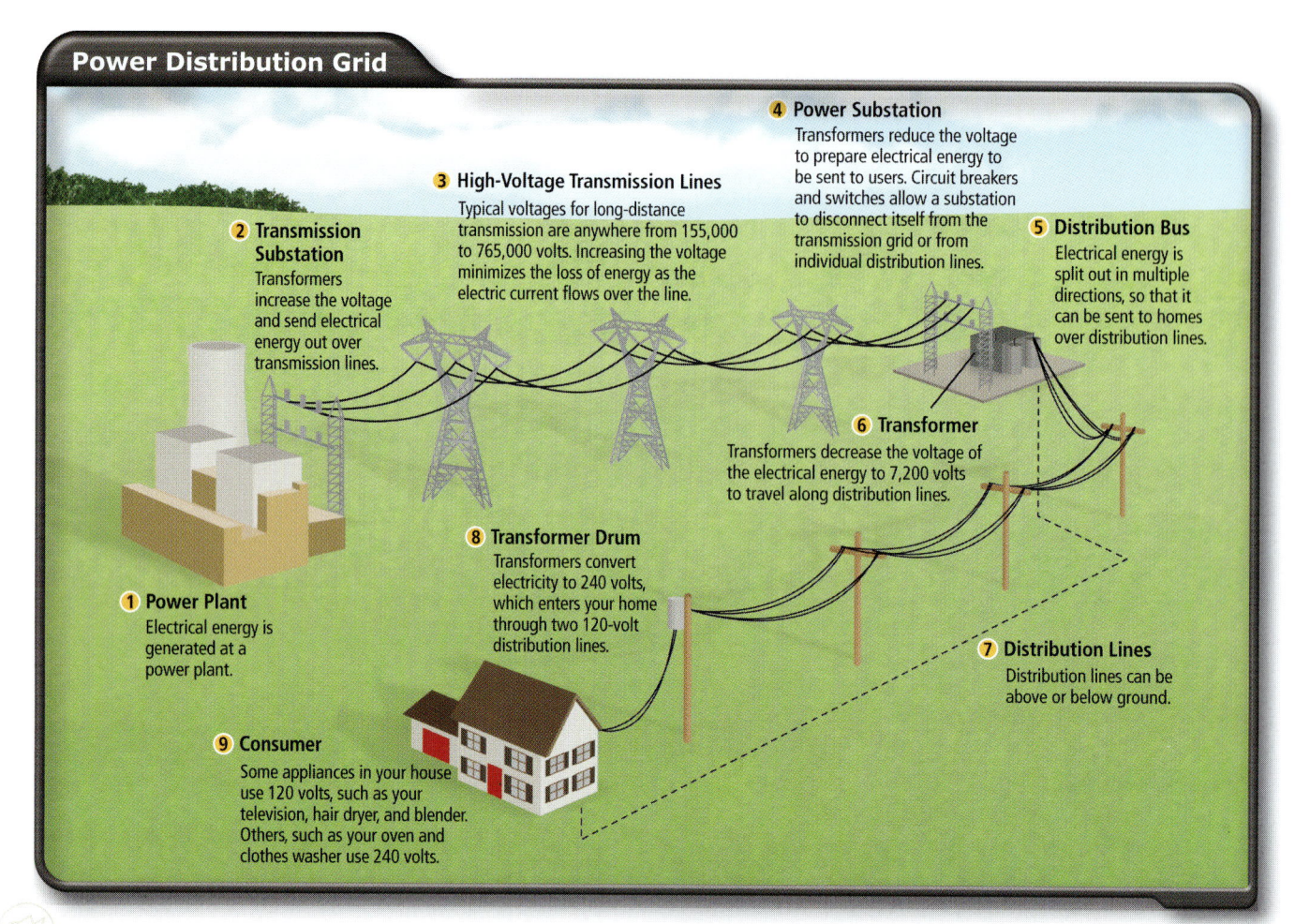

The Power Grid

The electrical energy produced at a power plant travels to your house through the **power distribution grid.** The power distribution grid is a system of facilities and structures that provide an organized and efficient way to deliver energy from the power plant to all of the people and things that need it. The diagram on this page illustrates how the power distribution grid works in North America.

The power that comes from a power plant in the United States is alternating current. The direction of the current alternates 60 times per second.

The main advantage of alternating current within the grid is that it provides a relatively easy way to change the voltage of the power using a device called a transformer. Transformers allow power companies to convert current electricity to very high voltages for transmission across power lines, then drop it back down to lower voltages for distribution and typical home use. The power that is available on transmission lines can reach 765,000 volts. The power that is available from a wall socket in the United States is 120-volt AC power.

Nuclear Power Plant

Marya/Wikimedia Commons

Team Highlight
Hiyam Añorve Garza, Tim West, Teacher Argonaut Melissa Hall, and Jackie Martin (L to R) examine the material used to simulate the sorbent in Larry's test facility.

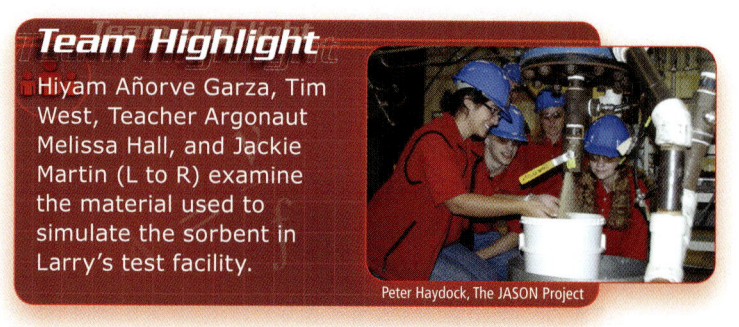

Peter Haydock, The JASON Project

60 • Operation: Infinite Potential   www.jason.org

Safety Features

Throughout the power distribution grid and in your own home, there are many key safety features to protect you. **Fuses** or **circuit breakers** can prevent fires. A fuse is a thin piece of wire that melts when too much current is sent through it, thereby breaking the circuit and stopping the current from flowing. A circuit breaker is a switch designed to automatically shut off when a circuit becomes overloaded. Meanwhile, a **ground fault circuit interrupter** (GFCI) outlet continually detects how much electric current is flowing. It shuts off the power whenever it detects any "leak" in the circuit that could cause an electrocution.

Fast Fact

On occasion, the power grid can experience brownouts or blackouts that disrupt power. A brownout is a temporary reduction in power, whereas a blackout is a temporary loss of power altogether. Brownouts and blackouts can occur due to grid damage caused by things such as storms. In other instances, a power company will actually decrease the supply of power due to excess demand.

The Modern Portfolio: Nuclear

Although the term *nuclear energy* may suggest some radical, alternative way of producing electrical energy, the overall process is similar to what occurs in coal-fired power plants. It involves creating steam, spinning a turbine, and transferring this mechanical energy to an electrical generator.

At a nuclear-fueled station, the thermal energy is produced by a reaction that occurs in the reactor core. Here, in a process called **nuclear fission**, atoms of a radioactive material, such as uranium, are split apart to release energy. During this breakdown, matter is transformed into a tremendous amount of thermal energy.

This thermal energy is used to heat water and create steam. This steam is used to spin the turbine blades. The steam is then cooled and condenses back into water. The water returns to the heat exchanger, and the cycle repeats.

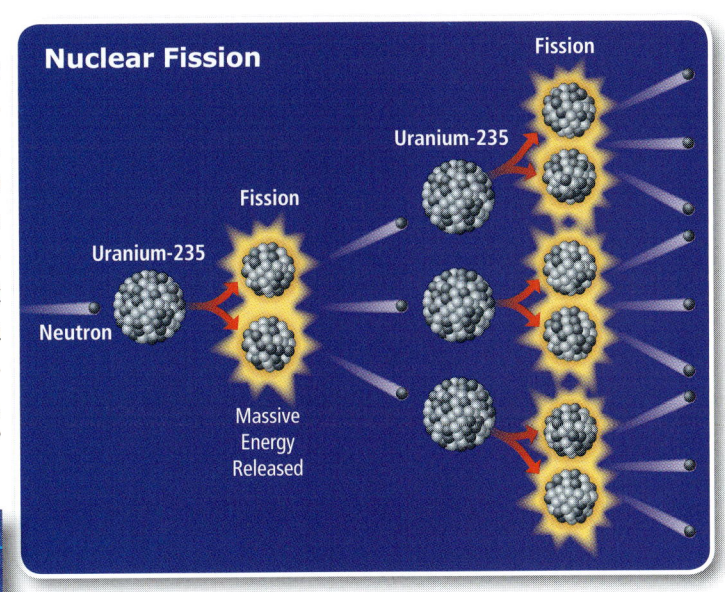

Nuclear Fission

Neutron — Uranium-235 — Fission — Massive Energy Released — Uranium-235 — Fission

Advantages and Limitations

Nuclear energy has several advantages. For one, a huge amount of energy can be generated from just a small amount of uranium. One pellet of uranium fuel, a few centimeters long, may produce as much energy as 150 gallons of oil. Also, nuclear energy does not produce the pollution that burning fossil fuels does. Splitting atoms produces no carbon-based pollution.

However, there are limitations. Spent fuel and the products of its breakdown are radioactive, which can have harmful effects on living organisms. In addition, some of this waste remains radioactive for thousands of years and must be safely stored. Continued availability of space for this storage along with long-term security and management are uncertain. Finally, uranium, is a **non-renewable** energy source which means it cannot be replenished once it is used. The mining for uranium can also disrupt natural ecosystems.

Nuclear fission (nuclear energy) ▶ **Boiling water** (thermal energy) ▶ **Steam** (mechanical energy) ▶ **Turbine** (mechanical energy) ▶ **Generator** (mechanical energy) ▶ **Current** (electrical energy)

The Modern Portfolio: Hydroelectric

Another way people obtain energy is through the use of flowing water. **Hydroelectric energy** is one of the oldest sources of energy. Hydroelectric energy is generated when water spins a turbine. Thousands of years ago, water was used to turn a paddle wheel to grind grain, like a windmill. Actual hydroelectric power plants that produce electrical energy have been in use in the United States since 1882.

Hydroelectric Systems

There are two main types of hydroelectric systems. In a "run-of-the-river system," the force of the water's current provides the energy to power the turbine. In a "storage system," water is collected in reservoirs created by dams, then released when needed. Hoover Dam in the United States and the Three Gorges Dam on the Yangtze River in China are examples of hydroelectric storage systems. When Hoover Dam was constructed, Lake Mead was formed as a result, and became the storage reservoir for the hydroelectric power plant.

Water (mechanical energy) ➡ **Turbine** (mechanical energy) ➡ **Generator** (mechanical energy) ➡ **Current** (electrical energy)

Advantages and Limitations

Many feel that hydroelectric energy holds great promise as an energy source for our future. Unlike fossil fuels or nuclear energy, hydroelectric power is **renewable**; water is recycled on the planet through the water cycle.

On the other hand, some believe that hydroelectric power plants are disruptive to the ecosystem. In the Columbia River, for instance, a series of dams get in the way of salmon trying to swim upstream to spawn. Different solutions are being offered to remedy the situation. For example, engineers have added "fish ladders" to the system which help the salmon "step up" the dam to the spawning grounds upstream. However, hydroelectric energy continues to have its critics. Disruptions caused to the land can cause negative effects for both human and animal life.

▲ Some fish rely on free movement up and down rivers in order to survive. Many fish live in salt water but spawn in freshwater rivers. Biologists build "fish ladders" to allow fish to access rivers without being hindered by dams.

▼ Hoover Dam transforms the mechanical energy of moving water into electrical energy. Working at full capacity this power plant can produce enough electricity to power a city of 750,000 people.

Lab 4

Water Wheel

The coal-fired power plants that Larry's research helps to improve use turbines that power generators. Hydroelectric energy uses the same principles to generate electrical energy. However, instead of utilizing the mechanical energy of steam to spin a turbine, hydroelectric energy spins turbines using the mechanical energy provided by water. A water wheel, set in motion, is driven by the flow of water through its paddles.

As you might imagine, the efficiency of the entire process is dependent upon the design of the wheel. Wheels that harness more of the water's energy can meet higher energy demands. In this activity, you will have the opportunity to explore water wheel design. You will construct a simple water wheel. From your observations, you will suggest and evaluate new designs.

Materials
- Lab 4 Data Sheet
- several pieces of rotelle (wagon-wheel) pasta
- paper clips
- 4 cups
- water-proof clay
- water
- materials provided by instructor

Lab Prep

1. Use print and online resources to learn about the history of water wheels. How were water wheels used by ancient cultures? What tasks did they accomplish? How are historic water wheels different from the turbines found in today's hydroelectric plants?

2. Roll out a thin strip of waterproof clay. Firmly press this strip along the outer rim of a piece of rotelle (wagon-wheel) pasta. Make sure that the rim is completely covered with a thick layer of clay.

3. Along the length of the clay, insert materials provided by your instructor to form a pattern of paddle-like extensions.

4. Open and straighten a paper clip.

5. Insert the straightened paper clip into the center of the pasta wheel so that the paper clip acts as an axle.

6. Use two lumps of clay to anchor both ends of the axle to the rim of your wide-mouth cup. The wheel should be positioned over the center of the cup. Spin the wheel. Adjust as needed to ensure that the wheel rotates freely.

Make Observations

1. Fill the other cup with water. Carefully pour the water onto the paddles of your water wheel. What do you observe?

2. Suppose you were to increase the height from which the water was poured. How might that affect the spin? Create a hypothesis. Then, test your hypothesis. Explain your observations in terms of the potential energy content of the poured water.

3. Compose a list of factors that might affect the efficiency of the observed energy transformation. (When evaluating wheel efficiency, why do you think that it is critical to maintain the same height from which the water is poured?)

4. Select one of the listed factors, and explain how you would measure its effect on the efficiency of the energy transfer and transformation. With your instructor's approval, create a new wheel design to improve efficiency of the transformation. Now compare your new design to your first design.

5. Is the new wheel more efficient? Explain. Can you further improve its operation? If so, how?

6. Once again, with your instructor's approval, create a new design. Is the new design more efficient? Explain.

7. **Journal Question** How might using a denser liquid in place of water affect the wheel's movement and how could this relate to energy generation?

The Modern Portfolio: Fossil Fuels

Fossil fuels are by far the most common energy source used in the world today. They are called fossil fuels because they come from the remains of once living plants and animals. These plants and animals died long ago, and instead of decomposing fully, they were trapped under layers of sediment. Over time, the heat and the pressure from the top layers of sediment caused the plants and animals to break down into simple forms of **hydrocarbons**. Hydrocarbons are organic compounds that consist only of hydrogen and carbon atoms.

Types of Hydrocarbons

Fossil fuels may be divided into three general categories—coal, oil, and natural gas. **Coal** is a black or brownish-black rock formed from the remains of plant life.

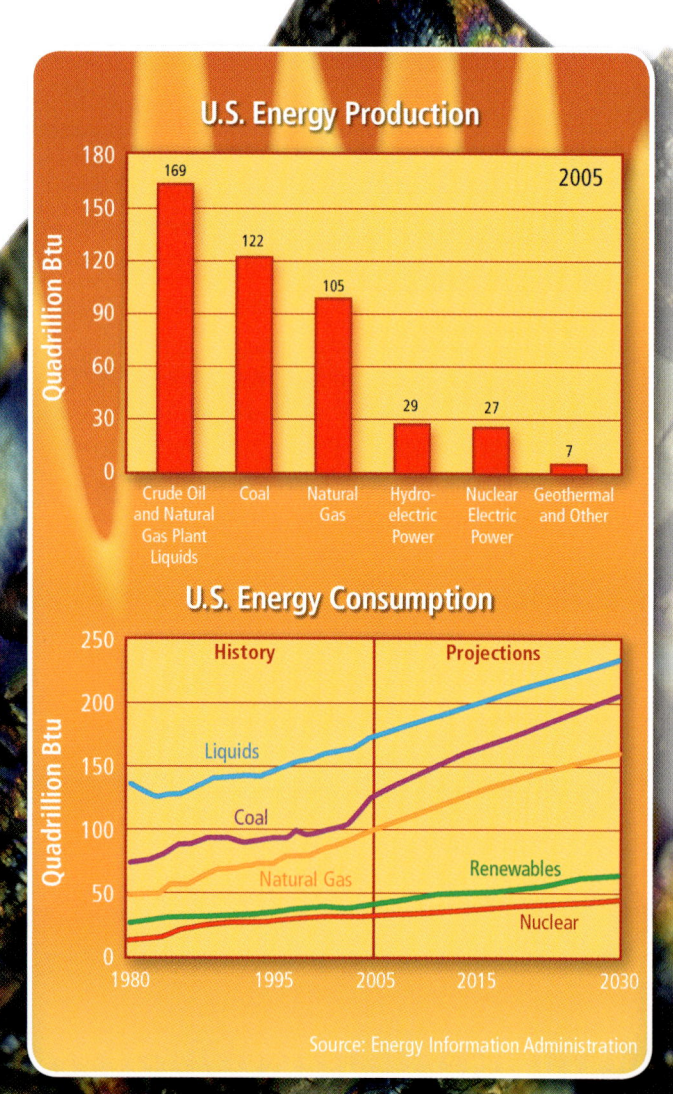

Oil is a yellow to black liquid formed from the remains of animals and plants that lived in a marine environment. Crude oil is the name of the substance that is removed from the ground. Crude oil can be sent to a refinery and made into other petroleum products, such as gasoline, kerosene, propane, and diesel fuel. A wide variety of other products are also made from petroleum—everything from tires and plastic products to ink and deodorant.

Natural gas is a colorless and odorless gas formed from the remains of animals and plants. Natural gas is a popular fuel choice because it burns cleaner and more uniformly than other fossil fuels.

Team Highlight
Creating a model of a generator, Hiyam Añorve Garza and Jackie Martin investigate its components and the variables that determine its efficiency.

Fast Fact
Shell uses enhanced oil recovery technology (EOR), to help get more oil out of the ground. Depending on the type or 'heaviness' of the oil, one of three EOR techniques are used: reducing the oil's viscosity by heating it with steam; injecting pressurized gas into the wells to 'push' the oil out; or, injecting chemicals that work like washing detergents to loosen the oil and help it flow. These techniques are helping recover oil that would otherwise be left behind in the ground.

U.S. Energy Production (2005, Quadrillion Btu)
- Crude Oil and Natural Gas Plant Liquids: 169
- Coal: 122
- Natural Gas: 105
- Hydroelectric Power: 29
- Nuclear Electric Power: 27
- Geothermal and Other: 7

U.S. Energy Consumption — History and Projections (1980–2030): Liquids, Coal, Natural Gas, Renewables, Nuclear.

Source: Energy Information Administration

64 • Operation: Infinite Potential   www.jason.org

Advantages and Limitations

One of the most significant benefits of fossil fuels today are the associated costs. Coal, oil, and natural gas have been an abundant resource for years, and are relatively inexpensive to drill or mine. Additionally, over the years, society has invested in an infrastructure that is designed to utilize fossil fuels. This includes cars, planes, factories, and homes. Changing this entire infrastructure to use another energy source would require another big investment.

A limitation to fossil fuels is that they are non-renewable and are becoming more scarce. Pollution also poses a significant challenge. When these hydrocarbons are burned, they produce carbon dioxide and other waste products. Adding large amounts of carbon dioxide into the atmosphere can have severe consequences. For instance, carbon dioxide can contribute to **global warming**, an increase in the overall temperature of Earth.

Solutions

Larry Shadle and his colleagues at the National Energy Technology Laboratory in Morgantown, West Virginia are working hard to minimize the effects of carbon emissions from burning coal. Scientists like Larry have figured out a way to use a chemical sorbent to remove carbon dioxide so that it is not released into the atmosphere. This is not easy, as coal in a power plant is burning 24 hours a day. The sorbent must be mixed in with the coal in just the right amount and at just the right time. Larry is proposing methods to mix the coal emissions with the sorbent in just the right amount, which would capture most of the carbon dioxide before it is released into the atmosphere. This technology is critically important, as it can allow us to use abundant coal resources for our energy needs now, while other forms of energy, such as solar, wind, or biofuels, are being further developed.

International Connection

LOCATION North Sea, United Kingdom

Monotowers

For decades, the North Sea has been a fertile source of oil and gas. However, as the fields reach depletion, remaining hydrocarbons are found in ever-smaller deposits. The development of monotower platforms by Shell has made it possible to recover this gas cost effectively and more cleanly in the southern North Sea.

In 2006, the Cutter Monotower began operating in the North Sea, United Kingdom. It was the world's first offshore natural gas production platform powered by wind and sunlight. Onboard the single-leg monotowers are solar panels and wind turbines. By using these inexhaustible energy sources, the natural gas is being extracted without producing any emissions. In fact, the platforms run on just 1.2 kW of power, less than it takes to boil a kettle of water. That compares to as much as 30 kW—supplied by cable, diesel generator, or gas engine—for a typical unmanned platform that provides the same production facilities.

Monotowers allow oil companies to have access to small pockets of natural gas in beds under the North Sea. Without monotowers, these natural gas fields would be uneconomical to drill using traditional techniques. The monotowers cost less than a third of most designs used a decade ago, and are cheaper to operate. Gas production is monitored and controlled from operation centers onshore, reducing safety risks. Also, maintenance visits are needed only once every two years, contributing to its low operation costs.

Field Assignment

Don't Leave Footprints

Recall that your mission is to *investigate how we currently meet our energy and power needs*. Now that you have been fully briefed, it is time to evaluate the effectiveness of our power-generating facilities and look at some of the emissions-control technology that they use.

Dr. Shadle and his team of researchers are focused on designing and implementing innovative emissions and carbon-control technology. Based on your analysis of Dr. Shadle's data, you will determine if he and his team are making a difference.

By employing carbon-control technology, coal-fired plants can reduce their carbon emissions and therefore reduce their **carbon footprint**. Reducing one's carbon footprint is a popular term these days—and so it should be. Countries around the world are beginning to monitor and promote strategies that reduce their carbon footprints in an attempt to curb global warming. Using secondary sources, you will obtain carbon emission and population data from a variety of countries around the world. You will analyze this data and see where your country stands. The first step in improving your country's standing is up to you. Start with your local area and determine the main sources of carbon emissions. After that, recommend ways your local area can help reduce its carbon footprint.

Peter Haydock, The JASON Project

Objectives: To complete this mission, accomplish the following objectives:

- Perform a comparison analysis of released sulfur dioxide (SO_2), nitrogen oxides (NO_x), and carbon dioxide (CO_2) data obtained from several emissions-controlled coal-fired power plants.
- Evaluate the effectiveness of emissions-control technology used in coal-fired plants.
- Using secondary sources, collect data on carbon emissions for China, India, United States, Qatar, and three other countries of your choice.
- Rank countries from most to least carbon emissions by country and then per capita.
- Identify the main sources of carbon emissions in your local community.
- Propose strategies that can help reduce your community's carbon emissions.

Mission 3 Argonaut Field Assignment Video Join the National Argonauts as they learn about emissions-control technology with Dr. Larry Shadle.

Caution! When surveying your local community, always do so with a responsible adult.

Materials
- Mission 3 Field Assignment Data Sheet
- graph paper
- ruler
- computer with Internet access

Field Preparation

1. Download the Field Assignment Data Sheet from the JASON Mission Center.
2. Using Dr. Shadle's data, create a bar graph comparing the amount of SO_2, NO_x, and CO_2 produced in different coal-fired plants.
3. Compare similarities and differences between the two data sets.
4. Identify and describe ways power plants can control their emissions.
5. Identify and describe some of the incentives for coal-fired plants to reduce emissions.

Mission Challenge

Your challenge is to compare the carbon emissions of several different countries. Additionally, you will propose strategies that can be used to help reduce your carbon footprint, and that of your local community.

1. Describe the term *carbon footprint* and explain how it relates to carbon emissions.
2. From the JMC, obtain links to current carbon emissions and carbon emissions per capita data for China, India, United States, Qatar, and three other countries of your choice.
3. Construct a bar graph to help organize your data and discuss your results.
4. Using your knowledge of how we currently meet our energy and power needs, identify the main sources of carbon emissions in your local community.
5. Propose a plan that will reduce the amount of carbon emissions in your local community.
6. Use an Internet carbon footprint calculator to determine your personal carbon footprint.

Mission Debrief

1. Explain why some countries have very high carbon emissions but overall low emissions per capita. Does their per capita emissions justify their high levels of carbon emissions? Explain.
2. Would you or your parents pay double for energy bills if the energy generation was guaranteed to be emissions controlled? Why or why not?
3. Identify some of the alternatives to fossil fuels that would be more emissions-friendly.

Journal Question Predict the role that fossil fuels will play as an energy resource in the future. Explain your reasoning.

Comparison of Coal-Fired Power Plants Using a Variety of Emissions Control Technologies

pollutants from coal-fired power plants	conventional combustor (with steam turbine and advanced pollution controls)	fluidized bed combustor (with steam turbine and nitrogen catalytic converters)	pressurized fluidized bed combustor (with gas and steam turbines but no nitrogen catalytic converters)	advanced gasifier (with gas and steam turbines and acid gas removal system)
Sulfur Dioxide (SO_2) lb / MW•h	2.0	3.9	1.8	0.7
Nitrogen Oxides (NO_x) lb / MW•h	1.6	1.0	1.7 - 2.6	0.8
Carbon Dioxide (CO_2) lb / MW•h	2000	1920	1760	1760

Connections

Culture

Appalachia

"*I'm a-fixin' to eat a mess o' flannel cakes.*" For many of us, this is probably a most unfamiliar statement. However, for some living in regions like the Blue Ridge Mountains in West Virginia, it may mean you are sitting down to a delicious breakfast of pancakes! If so, you may even be able to get some syrup from the "sugar tree."

In Appalachia, as in many regions throughout the world, unique sayings and speech patterns have evolved that define the area. But where do such expressions come from? And how do they emerge?

Dialects often emerge when individuals within a region are geographically isolated by landforms, such as mountains, limiting their exchanges with those outside the region and forming a close knit, insulated community. As a result of this relative seclusion from the "mainstream," such cultures can evolve a distinct and local form of a language called a "dialect."

Dialects, like languages, are not fixed. Over time, some words remain, new words are added, other words are blended or dropped, verb tenses and pronunciations are changed, and even the way in which sentences are assembled transform. Everyone, by definition, has a dialect. It's just that we are often not aware of it because we are surrounded by others who speak the same way we do.

So, where did the Appalachian dialects come from? In the 1600s, a large group of European settlers moved westward into the southeastern region of the Appalachian Mountains. In this harsh landscape, they joined the Native American population that had already settled in the mountainous uplands. Over time, this land would be included in the states of West Virginia, Kentucky, Tennessee, Georgia, and Alabama.

Most of the early Appalachian settlers could trace their roots back to the British Isles. They sailed from places such as England, Ireland, Wales, and Scotland. Even though these early settlers came from different regions of Europe, they all shared a somewhat common language—English. As they struggled within the unfamiliar frontier, their language evolved. Many words

---

How do you pronounce the word *Appalachian*? To some, especially those living in the northern half of the United States, it is often pronounced as "Appa-lay-shun." To most others, including the southern inhabitants of the United States, it is pronounced more like "Appalatch-in." Which one is correct?

According to most dictionaries, they are both acceptable ways of saying the word that arose from a group of Native Americans, the Apalachee tribe in Florida.

Blinked Milk

In Appalachian English, "blinked milk" is another way of saying that milk has soured, or gone bad. The term arose centuries ago, when it was believed that witches could cause all sorts of mishaps. It was thought that an evil eye—a blink from a witch—could sour fresh milk. That is how the term entered the regional dialect.

---

68 • Operation: Infinite Potential   www.jason.org

were retained from their native lands. But other common words were added, changed, and deleted over time. The result was the formation of unique dialects that are now known as part of Appalachian English.

In addition to developing distinct dialects, the isolated Appalachian regions also cultivated their own style of music. Like the language, this early music had roots in the Irish, English, and Scottish cultures.

The songs were mostly ballads — narrative poems with a repeating section called a refrain. The lyrics often told tales of love, relationships, and murder. While some of the stories were "Americanized," others retained their European setting and characters. In many of these remote settings, it was more often the women who sang these ballads. Assuming a nasal quality to their voice, they would sing unaccompanied. In contrast, musical instruments were played mostly by men. Their preferred instrument was the fiddle — a violin modified for faster play. Melody lines repeated over and over again, which lead to a distinct type of bowing that could be heard as the accompaniment to contra and square dances.

The music continued to evolve with an African-American influence. Spirituals that involved group singing and less structured lyrics expanded the style. Perhaps the greatest change was rhythmic. Slaves fashioned an instrument based upon instruments they had used in various parts of Africa. It became known as the banjo.

While it is important to realize that not everyone in a given region speaks exactly the same way or listens to the same music, these nuances of language and culture help to give groups of people a regional character. Thanks to thoughtful research into the origins and evolution of language and culture, people have now learned to appreciate and honor this region's unique dialect and music.

From Dialect to Dialect

Check out these Appalachian phrases and words and their standard English language equivalents.

A-fixin'	About to
Flannel cake	Pancake
Sugar Tree	Sugar Maple Tree
Plum	Completely
Reckon	Suppose
A fur piece	A long way away
Nigh way	A short distance
Heared tell	Learned
Let the latch down in the barn	Take advantage
Afeared	Afraid
Ill	Bad tempered
Yonder	There
Whistle Pig	Groundhog
Sop	Gravy
I don't care	Yes, please

Your Turn

Listen to your friends and family. Are there any special words, pronunciations, or sayings that might be part of a local dialect? How could you organize and compare dialects?

Connections: Culture • 69

Mission 4
Energy Independence
The Quest for Sustainable Resources

> *"These days, as a scientist, you need to reach out, so biology alone will not solve the problem, or chemistry, or physics. You have to bring it all together and combine it."*
>
> —Dr. Martin Keller
> Director, BioEnergy Science Center
> Oak Ridge National Laboratory

Martin Keller

Martin Keller leads a team of scientists and technicians including biologists, geneticists, and computer scientists who are working to produce ethanol from renewable plant sources.

Meet the Researchers Video
Join Martin and his team as they work to increase the supply of ethanol from non-food crop sources like poplar trees and switchgrass.

Director, BioEnergy Science Center, ORNL

Read more about Martin online in the JASON Mission Center.

Photos above (left to right): Andrew Henderson, NGS; Sebastian M/Wikimedia Commons; Joe Scherschel, NGS; Christian Mehlführer/ Wikimedia Commons; Peter Haydock, The JASON Project; Ian Martin, NGS; U.S. Air Force; Wolfgang Staudt/Wikimedia Commons

Your Mission...

Evaluate the future role of alternative energy resources.

To accomplish this mission successfully, you will need to

- Understand the differences between renewable, non-renewable, and inexhaustible energy sources.
- Recognize the importance of developing a diverse energy portfolio.
- Evaluate several alternate energy sources for their advantages and limitations.
- Discover the global potential of biofuels.
- Investigate alternate electrical production through solar, wind, and geothermal sources.

Mystery Connection

ALERT!

To: JASON Energy Consulting Team

A city in your region has been completely destroyed by a natural disaster. It has been determined that the energy grid and power plants have been damaged beyond repair. The government has decided that the city will be rebuilt using only renewable and inexhaustible energy sources.

Based on your understanding of renewable and inexhaustible resources, develop a plan for the local government. Design a new energy strategy for this city from the ground up. In your plan, include options for energy production and distribution in the region, and discuss the advantages and limitations of your choices.

Mission 4: **Energy Independence**—The Quest for Sustainable Resources • **71**

The Quest for Sustainable Resources

"Okay, who wants to go first?" The frigid Tennessee air remains motionless. No one replies. As the Argos approach the pen, the animals become interested in this odd group of humans. Called alpacas, the fleece-covered grazers resemble llamas. Curious, they walk toward their two-legged visitors.

The alpacas stare wide-eyed. Perhaps they wonder if the Argos will bring treats.

"Careful, one might spit," warns Dr. Martin Keller of the Oak Ridge National Laboratory (ORNL). He continues, "Who wants to take the first sample?"

The Argos, smiling uncomfortably, look at each other. Graciously, the Argos volunteer each other for the task that lies ahead—and what a task it is!

Dr. Keller's research focuses on the production of biofuels. Although a good deal of his scientific work occurs in the lab, he also goes into the field to collect living specimens. The kinds of organisms he studies are microbes that live in the scat, or droppings, of grazing animals or within the guts of wood-boring insects.

"Look! I think there's a fresh sample in the making!" announces an Argo.

Using a scoop fashioned from a tin cup fixed to the end of a stick, they collect fresh samples of scat. Still warm, the droppings are placed in containers and readied for analysis in Dr. Keller's lab. Later in the day, the samples will be brought to the lab.

The scientists are most interested in finding the enzymes from bacteria that have passed through the digestive tract of an alpaca. Using high-powered microscopes, these microbiologists search for cellulose-digesting bacteria that transform plant material into simple sugars that can be easily used for biofuel production. Perhaps these samples will contain microbes whose genes hold the key.

Mission 4 Briefing Video
Prepare for your mission by viewing this briefing on your objectives. Learn about alternative energy sources like biofuel, solar, and geothermal resources.

Mission Briefing

Evaluating Our Energy Portfolio

Imagine a world where non-renewable resources, such as coal and oil, run out! It may not be in our lifetime, but there will be a time when this will happen. What can we do now to prepare for this future event? We must research new sources of energy and take another look at **renewable** and **inexhaustible** sources that were once considered "alternative," such as sunlight and wind. As we prepare for a "greener" tomorrow, we must carefully weigh the advantages and limitations of each source.

72 • Operation: Infinite Potential   www.jason.org

Currently, renewable and inexhaustible sources of energy account for only about seven percent of our total world energy usage. By increasing the use of these sources, we could decrease our dependence on **non-renewable** sources, such as coal and oil, and extend the amount of time these resources would be available. We may even help the environment along the way!

Energy Portfolio Options

How can you decide which energy sources are viable options for your local area? We are living in a world where science and technology give us more energy options than ever. Wind, sunlight, geothermal, waves, tides, biofuels, nuclear, and fuel cells are just some of the energy sources you could have access to when restructuring your energy portfolio.

Energy Portfolio Considerations

It is likely that no single energy source will solve the problem on its own. Therefore, think about how you could use several different energy sources within your region. What factors should you consider?

First, you need to assess the current and future demand for energy in your area.

Second, you need to become educated about current and future energy resources that are, or could be, available and appropriate for your location.

Finally, consider the impact each energy source has on the planet. For instance, can it help reduce our carbon footprint? A **carbon footprint** is a measure of human impact based on the emission of carbon-based greenhouse gases, such as carbon dioxide. Keep these considerations in mind as you read about each energy source described on the following pages.

When we rethink our energy options, we come up with innovative ways to capture and utilize energy to suit our everyday needs.

Join the Team

Dr. Martin Keller, from the BioEnergy Science Center (BESC) at Oak Ridge National Laboratory (ORNL), and Argonauts Hannah Zierden, Joey Botros, Hiyam Añorve Garza, and Melissa Hall (L to R) meet at Dr. Keller's farm. They discuss the results of their research in biofuel production. Dr. Keller and his staff of over 300 scientists and technicians are investigating the production of ethanol from nonfood-based crops. Poplar trees and switchgrass are two major plants being researched at ORNL as a potential source for biofuels. The Argonauts' original research found bacteria in the gut of a grub that may lead to more efficient production of ethanol from poplar trees.

Mission 4: Energy Independence—The Quest for Sustainable Resources • 73

Solar Hot Water Systems

Passive Solar System | **Active Solar System**

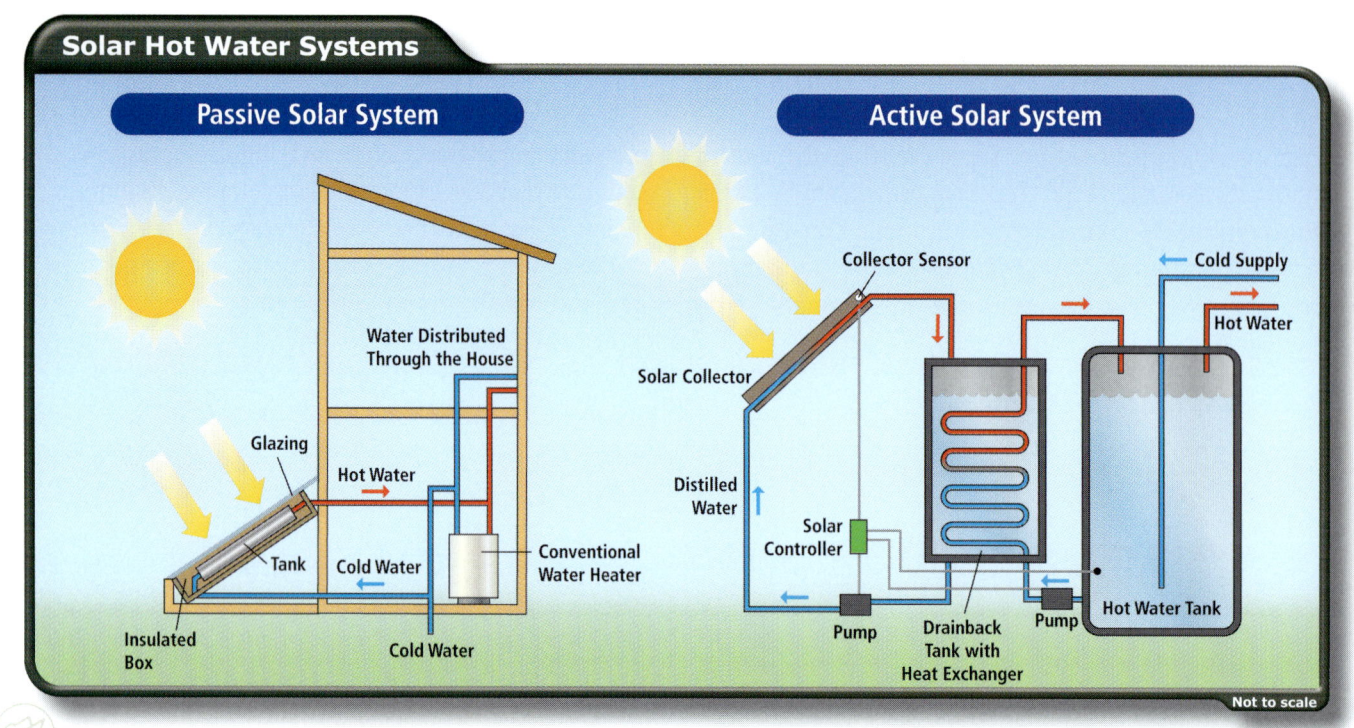

Collecting the Sun's Energy

In many areas of the world, such as in the Mojave Desert and southern Spain, solar energy generates heat and electricity for tens of thousands of homes. Powering TVs, computers, lights, and heaters, the sun meets much of people's everyday needs in these regions.

Solar heating is another way to heat water and your home. Collecting solar energy for heating can be accomplished using two types of systems—passive and active.

Passive and Active Solar Heating

Passive solar heating systems simply collect electromagnetic energy without the use of any moving mechanical parts, such as pumps and fans. Have you ever stood in the sun in a black t-shirt? What did you notice? The sun's energy was passively absorbed by the dark shirt and your body. This absorbed energy is transformed into thermal energy, which then warms your shirt and your body.

Fast Fact
During the Little Ice Age (around 1500–1750 A.D.), French and English farmers set up fruit walls in their fields to collect electromagnetic energy to keep their plants warm. When exposed to sunlight, stone or brick walls quickly warm and retain heat. In cooler climates, planting trees and crops near or on these walls would warm the plants, increase their growth, and extend the growing season.

Architects and engineers design walls, windows, floors, and roofs that can passively collect and transform the sun's energy into thermal energy. They do this by facing more windows towards the sun in cooler climates. This allows more electromagnetic energy onto floors and walls that then absorb the energy and transform it into heat. Dark stones or tiles are often used to maximize this energy transformation. This way homes and buildings can cut down on energy bills with minimal environmental impact.

The difference between passive and **active solar heating** systems is that active systems collect and move solar energy with the use of moving mechanical parts, such as a pump or a fan. The addition of a pump or a fan increases the efficiency of the collection, storage, and distribution of thermal energy within the system. Often, thermal energy from active solar systems can be stored in rock bins or in water tanks. This energy can then be used later when the electromagnetic energy from the sun is no longer available. So, when the sun is not shining, the solar energy is still available for use!

Solar Power Plants

Scientists are also now able to transform solar energy into electrical energy. At a solar power plant, the sun's energy is first collected and concentrated by reflective troughs, dishes, or towers. This energy is then used to heat a fluid, such as water, which produces steam. The steam is then used to spin a turbine which turns a generator to produce electrical energy.

Lab 1

Wind Power

Although most of the world's electrical energy supply is produced in fossil fuel-powered plants, Martin Keller believes in a diverse energy portfolio. Advances in technology have lead to increased usage of biofuels and wind turbines. Wind turbines depend upon the natural movement of air to spin a generator's shaft. Unlike the burning of carbon-based fuel, the harnessing of wind to generate electrical energy does not release greenhouse gas emissions.

In this activity, you will construct a basic wind turbine. After observing it in action, you will analyze the design and propose improvements to its generating capabilities. Then, you will test your suggestions to determine the most efficient model.

Harnessing the power of the wind allows us to use this inexhaustible energy source.

Materials
- Lab 1 Data Sheet
- 1.5-volt motor
- multimeter
- tape
- pushpin
- scissors
- fan
- cork or rubber stoppers
- 2 connecting wires (with alligator clips)

Lab Prep

1. Carefully use the pushpin to place two holes in the center of each end of the stopper.

2. Insert the shaft of the motor into the hole made at one end of the stopper. Make sure that the attachment is secure.

3. Download the Lab 1 Data Sheet. Cut out the pinwheel template according to the instructions provided.

4. Bend the dotted tip of each of the four sections of the cut paper to the center of the square. The bent paper should form a pinwheel design, which is now your turbine.

5. Carefully insert a pushpin through each of the four blade dots. Continue inserting the pushpin into the central dot. Then, insert the end of the pushpin into the free end of the stopper.

6. Use tape to attach the motor to the edge of a table. Make sure that the turbine spins freely.

7. Use the alligator clip connectors to attach the motor leads to the multimeter. Set the multimeter to measure DC voltage.

Make Observations

1. Blow on the turbine. How does this affect the measured voltage? Does varying the force of the wind affect the measured voltage? Explain any relationship you observe.

2. As a class, review any safety concerns associated with an electric fan. Your instructor will turn on one or more classroom fans. Observe how the moving air affects your turbine.

3. Observe a turbine that is in the direct path of the air flow from the fan and one that is not. How does the position of the turbine affect the voltage generated?

4. Consider the setup you have constructed. Make a list of different factors that might affect the efficiency of this energy transfer and transformation. How can the design be improved? Which of these factors can you explore by extending your inquiry?

5. Discuss your inquiry extension with your instructor. With your instructor's approval, explore how changes in design can improve the amount of voltage generated.

7. **Journal Question** Compare and contrast a wind turbine that generates electrical energy with a windmill that is constructed to grind corn or pump water. Use the terms *transfer* and *transformation* in your answer.

Mission 4: Energy Independence—The Quest for Sustainable Resources

Photovoltaics

You do not necessarily need an entire solar power plant to transform the sun's energy into electrical energy. In fact, most often when we talk about generating electrical energy from electromagnetic energy, we are referring to using photovoltaic systems.

Photovoltaic (PV) technology focuses on transforming electromagnetic energy from the sun into electrical energy. You may be familiar with the small, light-sensitive strip that powers some calculators. Or maybe you have seen a solar panel on a road sign or a garden light. Most PV technologies use similar concepts in order to produce electrical energy.

Most photovoltaic systems use a photovoltaic array. A **photovoltaic array** consists of multiple photovoltaic cells linked together to generate a higher energy output.

Today, products such as PV plastics, fibers, and paints, called thin-film photovoltaics, are changing the way we look at and think about electromagnetic energy collection.

Advantages and Limitations

Solar power is popular because it comes from a free and inexhaustible source. In addition, once installed, solar energy systems do not produce any air or water pollution. PV cells and arrays can also be designed to fit most size or shape requirements, making them extremely useful in different situations.

Major limitations of solar power include the initial cost of the equipment and installation. Large scale solar arrays and power plants also require large amounts of land. Additionally, the amount of electrical energy produced depends on the amount of sunlight available.

Layers of a PV Cell

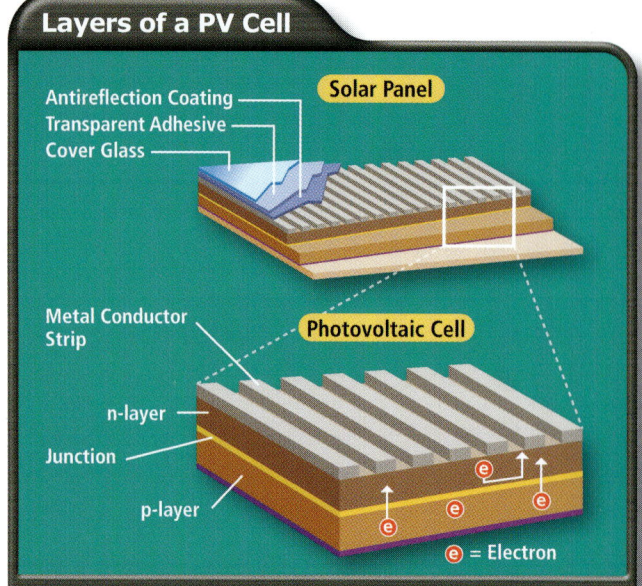

Solar Panel
- Antireflection Coating
- Transparent Adhesive
- Cover Glass

Photovoltaic Cell
- Metal Conductor Strip
- n-layer
- Junction
- p-layer

e = Electron

PV cells can transform electromagnetic energy into electrical energy because of the properties of two specially treated materials. When these materials are placed on top of one another, an electric field develops.

1. Sunlight passes through the top layer (n-layer) and strikes the bottom layer (p-layer).
2. The bottom p-layer of the PV cell is specially treated to release electrons when it absorbs a certain amount of electromagnetic energy.
3. Electrons released from the bottom p-layer follow the electric field through a junction and are collected by the top n-layer. This top n-layer has been specially treated to accept electrons.
4. Because of the direction of the electric field created by the junction of both materials, it is harder for electrons to return back through the junction to the bottom p-layer.
5. As more electromagnetic energy reaches the bottom p-layer, and as more electrons move to the top n-layer, a voltage builds.
6. A metal conductor strip creates a circuit that carries electrons away from the top n-layer back to the bottom p-layer. This is an easier pathway for electrons to return back to the bottom layer. This flow of electrons is the electrical energy that can be used to charge batteries and light lamps.

Not to scale

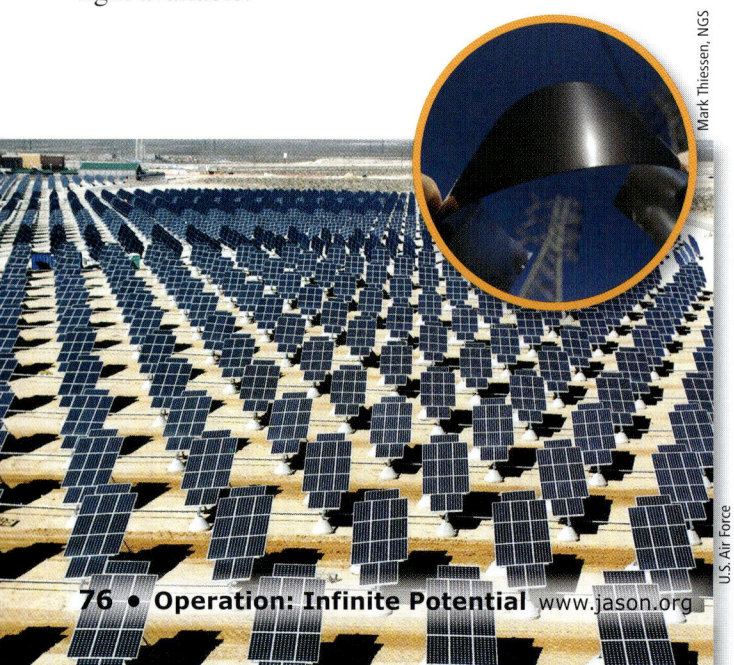

Individual photovoltaic cells are composed of silicon sheets. Linking these cells together creates a photovoltaic array, which produces levels of electrical energy high enough to be distributed to the power grid.

76 • Operation: Infinite Potential www.jason.org

Lab 2

Generating Hydrogen Gas

Although Dr. Keller is most interested in the production of biofuels from cellulose, there are other alternatives to fossil fuels. Hydrogen gas is one such alternative. Like gasoline and alcohol, it can be burned in a combustion engine. However, the combustion of hydrogen does not produce carbon dioxide or toxic air pollutants. Instead, it produces water vapor.

Hydrogen can be obtained through several chemical reactions. One process uses electrical energy to split water molecules into their basic components. Known as electrolysis, the process is used to generate hydrogen gas that can later be used as a cleaner fuel. In this activity, you will explore electrolysis and generate hydrogen gas.

Materials
- Lab 2 Data Sheet
- 2 connecting wires with alligator clips
- 9-volt battery
- 2 pencils with eraser and metal end removed
- tap water
- hand lens
- pencil sharpener
- masking tape
- clear plastic cup

Lab Prep

1. Carefully sharpen both ends of each pencil.
2. Fill a clear plastic cup about halfway with tap water.
3. Place the pencils in the cup so that one end of each pencil is submerged. Separate the points by about 2.5 cm. If needed, use tape to secure the pencils.
4. Use a hand lens to examine the sharpened pencil points above and below the water. Make a drawing of what you see.
5. Attach one alligator clip connector to each sharpened pencil point that is above water. Do not connect the free ends of these wires. For now, let them remain unattached.
6. Once again, use a hand lens to examine all four pencil points. Have their appearances changed? Explain.

Make Observations

1. Connect one of the free alligator clips to the negative terminal of the 9-volt battery. Connect the other free clip to the battery's positive terminal.
2. Examine the pencil points. What do you see? Do the appearances of the pencil points change? If so, how? Draw a picture of what you see.
3. Examine the points for several minutes. Compare and explain their appearance.
4. Research the chemical formula for water. What kinds of atoms are in a water molecule? With this new information, refine your answers in step 3.
5. Considering that electrolysis is an inefficient energy transformation, evaluate an advantages and limitations of using the electrolysis as an energy source to produce a fuel.

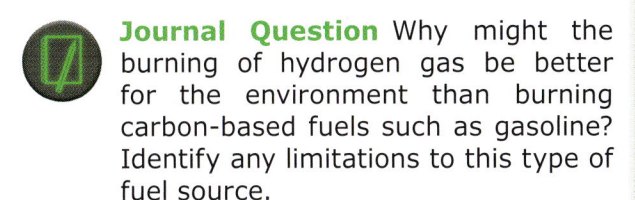

7. **Journal Question** Why might the burning of hydrogen gas be better for the environment than burning carbon-based fuels such as gasoline? Identify any limitations to this type of fuel source.

Mission 4: Energy Independence—The Quest for Sustainable Resources

Biofuels

As oil and gas supplies diminish, scientists like Dr. Martin Keller look for substitutes for these energy sources. Because plants and algae are the original sources of oil and gas, it made sense to start research with these renewable sources. **Biofuels** are a renewable, carbon-based energy source derived from biomass. **Ethanol** and biodiesel are among the biofuels that can be derived from these biological sources.

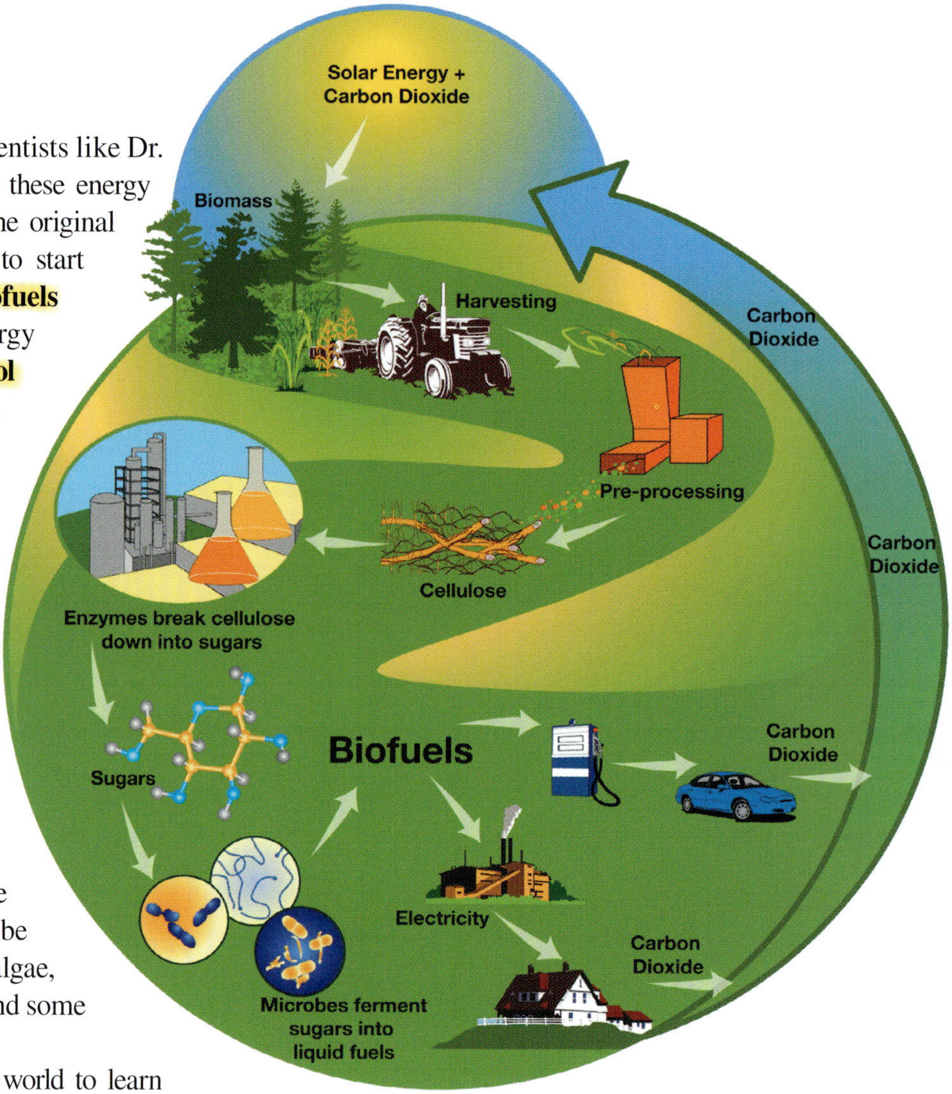

Manufacturing ethanol from biomass such as corn, switchgrass, and sugarcane is commercially viable today. Brazil runs most of its cars and trucks on ethanol derived from sugarcane. In North America and Australia, gasoline blends that contain 10 to 85 percent ethanol are used, creating cleaner-burning gasoline and stretching supplies.

Biomass can also be used to create diesel substitutes. These substitutes can be derived from seeds of plants, soybeans, algae, and canola oil. They can power trucks and some cars without any engine modifications.

Scientists often look to the biological world to learn how to produce biofuels. The digestive systems of plant-eating animals may contain huge populations of cellulose-digesting microbes. These simple organisms contain enzymes that break down cellulose into sugars. The sugars are then fed to yeast, a type of fungus, which produce ethanol. Dr. Keller's research group looks all over the world for new microbes to test, with the hope of finding enzymes that more efficiently break down cellulose.

Common Biomass Materials

| Trees and grasses (poplar, switchgrass, willow, silver maple, eastern cottonwood) | Agricultural crops (corn, sugarcane, bamboo, wheatgrass) | Garbage and municipal waste (paper, cardboard, yard waste) | Aquatic plants (algae, kelp) | Animal waste (produced by cows, chicken, pigs) |

78 • Operation: Infinite Potential   www.jason.org

Dr. Keller's group also looks for ways to manipulate the DNA of plants so that their cellulose is easier to digest. The perfect plant would be easily and completely digested by enzymes, resulting in sugars that could be fermented by yeast to make ethanol.

Advantages and Limitations

The combustion of biofuels produces no net carbon dioxide emissions. For example, if you rake up leaves from a tree in your backyard and send them to a power plant to be burned, the burning leaves at the power plant release carbon dioxide. However, the tree that dropped those leaves in the fall absorbed about the same amount of carbon dioxide in the spring when it made those leaves. Compared to coal, which takes millions of years to be replaced, the combustion and processing of biomass does not dramatically increase carbon levels in the atmosphere.

Though biofuels may provide a possible solution, they do have a few limitations. The use of biofuels has created an argument over food versus fuel. Agricultural land that was once used for food production is now being used to grow plants for energy. Also, burning biomass does generate some air pollution, such as nitrogen oxides and carbon monoxide. Finally, some biomass produces unpleasant odors when it is produced, refined, or burned.

Fast Fact

Tempura, a popular Japanese dish, is making big headlines overseas. In Japan, the leftover cooking oil used to fry this Japanese specialty has been used to power race cars, cruise ships, and even airplanes!

Team Highlight

Dr. Keller and Dr. Lee Gunter introduce the Argonauts to techniques geneticists use to make poplar trees grow with cell walls that will be easier to break down into biofuels. The source material is the first step in the production of ethanol. Other plants such as corn, sugarcane, and sugar beets are also used. However, these plants are more common as food crops and can cause food prices to rise if they are used as biofuels.

Mission 4: Energy Independence—The Quest for Sustainable Resources

Energy from the Wind

Throughout history, humans and animals have harnessed the energy of the wind. All over the world, sailboats and windmills can be found. Early windmills were also used to pump water and grind grain.

In the late 1800s, many parts of the world were starting to use electricity. In 1887, in Glasgow, Scotland, a small windmill was used for the first time to generate electricity for one cottage. As demand for electricity began to grow quickly around the world, interest in wind energy faded because wind technology could not keep up with the energy demand.

However, over the last two decades, attention on the environmental and economic effects of using inexhaustible resources has brought wind energy back into the picture. Today, wind is the fastest-growing segment of the energy industry.

At the heart of the wind energy industry is the wind turbine. A **wind turbine** is a large, fan-like structure used to capture wind energy. As the wind turns the rotors, a generator transforms the mechanical energy of the wind into electrical energy. **Wind farms** are groups of wind turbines that work together to produce higher amounts of electrical energy.

Advantages and Limitations

Like solar energy, the amount of electrical energy that can be produced from the wind in a given location is dependent on a few factors. The average wind speed and distance between the wind farm and the consumer must be considered, as well as the amount of time the wind blows.

However, wind energy produces no waste or carbon emissions, is inexhaustible, and, in the long run, is fairly inexpensive. Most, if not all, of the expenses for wind energy production come from the building and maintenance of the wind turbines and farms. The fuel is free, and there is an infinite supply of it.

Cray XT, aka Jaguar

Oak Ridge National Laboratory in Tennessee is home to Jaguar, the world's most powerful supercomputer dedicated to open science. Jaguar is capable of performing up to 1.64 quadrillion calculations a second (or 1.64 petaflops), making it up to 100,000 times faster than your average laptop. The speed and volume at which the Jaguar can process and calculate information provide an unprecedented potential for scientific investigation. In order to help further science, the Department of Energy's Office of Science makes Jaguar available to researchers throughout academia, industry, and government. This supercomputer gives scientists like Dr. Martin Keller the tools necessary to quickly analyze and process enormous amounts of data, which will ultimately aid him and his team of scientists in the search and discovery of new and innovative forms of energy.

Name: Jaguar

Peak performance: 1.64 petaflops (quadrillion calculations per second)

Number of processing cores: 182,000

Processor type: AMD quad-core Opteron™ 2.3 gigahertz

Memory: 362 terabytes

Memory bandwidth: 467 terabytes per second

Jaguar vs. Average Laptop Computer

	Jaguar Supercomputer	Average Laptop Computer
Speed	1.64 petaflops (i.e., 1.64 quadrillion calculations per second)	16 gigaflops (i.e., 16 billion calculations per second, or 0.000016 petaflops)
Number of processing cores	182,000	2
Processor type	AMD quad-core Opteron™ 2.3 gigahertz	Each core at 2 gigahertz
Memory	362 terabytes	2 gigabytes (0.002 terabytes)
Memory bandwidth	467 terabytes per second	25 megabytes per second (or 0.000025 terabytes per second)

Wind farms have a few limitations. They require large amounts of land. Air pressure changes caused by the blades of the turbines also have some negative effects on wild bird and bat populations, and could have effects on their migratory paths. Because a lot of wind energy comes from farms on coastal areas, there is also debate as to whether the farms could interfere with marine environments, including whale migration. Many people also claim that the visual impact of the turbines and farms makes them unappealing.

80 • Operation: Infinite Potential   www.jason.org

Lab 3

Biofuels: Into the Woods

In the field, Dr. Keller studies and collects various organisms. These organisms are the source of special proteins called enzymes. The enzymes that Dr. Keller seeks are those that play a role in the conversion of cellulose into starch and simple sugars to make ethanol.

In this activity, you will use satellite imagery to identify a local habitat that may support various cellulose-consuming organisms. You will plan a field trip to this habitat to study and develop a species list of organisms as you observe and record evidence of cellulose consumption.

Materials
- Lab 3 Data Sheet
- computer with Internet access and Google Earth™ installed
- field guides
- field glasses
- digital camera
- hand lens

⚠ Review safety precautions for field work with your instructor before beginning this lab.

Lab Prep

1. Research and create a list of local plants and organisms that feed on those plants. In what location and habitat would you expect to find each of these organisms? What evidence would support their presence and cellulose-eating behavior?

2. Launch Google Earth™ to explore the geography of your local surroundings. Identify any nearby forested areas, fields, grasslands, or other habitats that might support organisms that feed upon cellulose. From the list above, what organisms do you expect to find in these locations?

3. With your instructor, plan a field trip to your selected habitats for the purpose of locating evidence of organisms that obtain nutrients from cellulose.

Make Observations

1. At the site, update your map with features that are different from the satellite images. Plot the locations of the larger and dominant plants. Use a field guide to help you better identify all specimens encountered.

2. Throughout your trip, record and take pictures of the habitats.

3. Examine fallen trees and decaying stumps, leaves, or soft plant parts for evidence that suggests that this biomass is being consumed by animals or broken down by decomposers.

4. What clues might lead to the consumer's or decomposer's identity?

5. Back in the classroom, compare and contrast the species lists compiled by the different teams. Address any differences.

6. Based upon your investigation, what local species could be further studied to learn more about cellulose digestion and metabolism? Support your choices.

Journal Question Some biologists believe that the Endangered Species Act should protect the smallest organisms such as bacteria and fungi. Justify this position with respect to alternative energy resources such as biofuels.

Mission 4: Energy Independence—The Quest for Sustainable Resources

Waves and Tides

The motion of water contains a large amount of untapped energy. Have you ever tried to stand up on the beach while a big wave crashed into you? Or when you sit in a boat on a large body of water, what do you notice? The boat bobs up and down as the waves pass under the boat. Ocean energy generators can transform these water movements into electrical energy.

Fast Fact
Construction began on the first commercial wave energy farm in 2008, when three of the 28 Pelamis Wave Energy Converters (PWECs) were towed into position 5 km (3 mi) off the coast of northern Portugal. Measuring 140 m (460 ft) in length, each PWEC buoy spins internal generators, which can provide up to 2.7 GW•h (billion watt hours) of energy annually to the grid and could reduce CO_2 emissions by 2,000 tons each year. Upon completion, the 28 machines will provide electrical energy for up to 15,000 local residences.

Team Highlight
The Argonauts split wood and collect alpaca scat on Dr. Keller's farm. They are looking for sources of bacteria in the scat and beetle grubs that could be used to help break down cellulose into sugar.

Energy from Waves

Scientists and engineers are working on technology that can efficiently transform the mechanical energy found in waves into electrical energy. A variety of wave energy devices are currently in operation. One type utilizes the vertical displacement of the water column. As waves travel through the ocean, buoys move up and down. Engineers use this motion to oscillate a magnet through a coil of wires, which ultimately produces electrical energy. Engineers are also using wave energy on coastlines. Water from these waves compresses a column of air inside a structure that spins a turbine, which drives a generator to produce electrical energy.

Energy from Tides

Tides are the continuous rise and fall of Earth's ocean surface caused by gravitational forces between the sun, moon, and Earth. These interactions cause large amounts of ocean water to move into and away from shorelines every day. During high tide, ocean water moves toward the shore. Then, during low tide, ocean water moves away from the shore.

The Bay of Fundy in Canada has one of the highest vertical tidal ranges in the world. In some areas of the Bay, the water level can go up and down 17 m (56 ft) twice in a day.

Tidal electric generators transfer the mechanical energy of tides into electrical energy. Many tidal power systems use a tidal turbine. Tidal turbines resemble wind turbines, but are placed under water. As the tide flows into shore, the moving water pushes against the turbine blades, causing them to spin. The spinning turbine blade powers a generator, which produces electrical energy. As the tide changes directions and flows outward, the turbine continues to power the generator by spinning in the opposite direction.

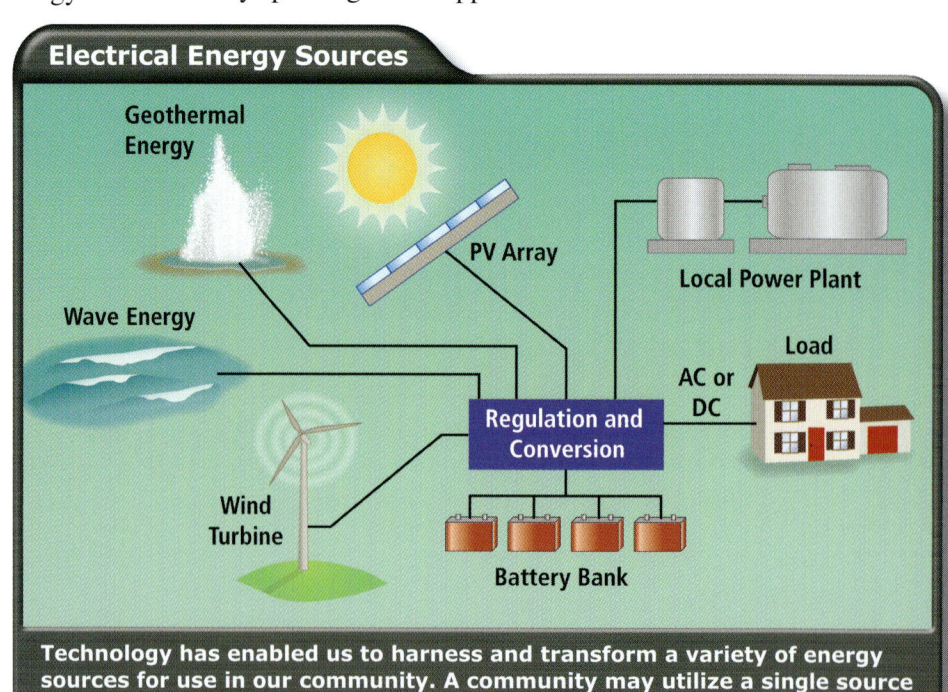

Electrical Energy Sources

Technology has enabled us to harness and transform a variety of energy sources for use in our community. A community may utilize a single source or combination of energy sources to supply energy to the grid.

82 • Operation: Infinite Potential   www.jason.org

Advantages and Limitations

Waves and tides are a free, reliable, and predictable resource for energy. They produce no emissions or pollutants and unlike fossil-fuel powered plants, some marine energy devices are not as visible from land.

Because generating energy from waves and tides is relatively new, many of the limitations are still unknown. How wave and tidal turbines may alter the ecosystem of a bay or coastline is still being evaluated.

Additionally, there could be some battles regarding rights and ownership. Generally, a country owns submerged land out to the end of the continental shelf. However, because that land is owned by the government, there is a question of whether ownership of the tidal farms could be privately held. That is a question that will need to be addressed in the future.

Currently, tidal technology involves trapping water at high tide and funneling the collected water through energy-generating turbines at low tide. This could have significant environmental impacts. Less intrusive designs, such as tidal turbines (above), may hold the key for a tidal energy future.

The Potential of Nuclear Fusion

Nuclear fusion is a process that naturally occurs within the sun. Within the extreme pressure and temperature of the sun's core, hydrogen nuclei are joined, or "fused" together to produce helium nuclei. During this process, a tiny bit of matter is transformed into an awesome amount of energy. Using nuclear fusion, one gram of matter (about the mass of a dollar bill) produces approximately the same amount of energy in approximately 20,000 tons of dynamite.

Advantages and Limitations

Even though nuclear fusion is still only in development, it promises to provide many potential advantages for the future. Fusion reactions might provide three to four times more energy than the fission reactions used in current nuclear power plants. Also, the fuel used for a fusion reaction is easily obtained from water, which is widely available. Additionally, fusion power plants do not produce greenhouse gases or particulate pollutant emissions. Helium, a harmless inert gas, is the primary byproduct of fusion reactions, which can be collected, stored, and used as additional fuel. Furthermore, compared to other energy sources, such as sunlight and wind, the area needed for a fusion power plant is much smaller.

The fusion process does pose some limitations, such as the resulting build-up of radioactivity. Additionally, at this time, it is not a cost-effective means of producing energy. We must thoroughly investigate any other limitations before nuclear fusion can be made commercially available.

International Connection

LOCATION Cardache, France

The first fusion experiments began in the United Kingdom in the 1930s. After World War II, there was renewed interest in harnessing the power of atomic fusion to make energy and use it in peaceful ways.

Around the world, laboratories in universities and government facilities began experimenting. A conference in Geneva in 1958, called "Atoms for Peace," was the beginning of global collaboration in the development of fusion as a viable energy option.

The United Kingdom has the largest fusion reactor test facility built to date called the Joint European Torus (JET) in Culham.

In 2006, a joint venture between the Russian Federation, Japan, China, India, South Korea, the United States, and the European Union was created to demonstrate that fusion technology could be used to produce electrical energy. Originally called the International Thermonuclear Experimental Reactor, the consortium is now just known as ITER.

Construction of the ITER research facility began at Cardarche in the South of France in 2008. Based on JET experiments, ITER estimates that a fusion generator can become a feasible power source by 2050.

Lab 4

Biofuels: Into the Lab

Dr. Martin Keller and his team are interested in learning how microbes, like yeast, produce ethanol. Through their field work around the world and lab work at Oak Ridge National Laboratory, they hope to discover natural substances that can breakdown cellulose into sugar. If these chemicals are discovered, then plants and plant material that would otherwise be discarded as waste can be processed into biofuel.

During fermentation, sugar (glucose) is broken down into ethanol and carbon dioxide gas. By measuring the volume of generated gas, you can evaluate the effectiveness of this chemical process. In this activity, you will evaluate fermentation scenarios based upon the observed production of carbon dioxide.

Materials
- Lab 4 Data Sheet
- freeze-dried baker's yeast
- sugar
- grass clippings
- cornstarch
- amylase
- tablespoon
- 6 8-oz. clear bottles
- 6 balloons
- graduated cylinder
- balance
- hand lens
- safety goggles
- stopwatch
- thermometer

Lab Prep

1. Review the formula for fermentation found on this page.
2. What is amylase? Where is it found? How might it be used to aid in the production of ethanol?
3. What is cellulose? How is cellulose different from starch? What type of organisms naturally break down cellulose? Where are they found?

Make Observations

1. Label six clear bottles A, B, C, D, E and F.
2. Put on your safety goggles. Do not add anything to container A. Add 1 tablespoon of sugar to container B. Add 1 tablespoon of cornstarch to containers C and D. Add one tablespoon of grass clippings to containers E and F.
3. Add 1 tablespoon of yeast (about 0.8 g) to each container.
4. To containers D and F, add 100 mg of amylase, as supplied by your instructor.
5. Have a balloon ready to cap each bottle. Add 100 mL of tap water at the temperature recommended by the yeast packaging to each bottle. As soon as water is added, stretch a balloon over the containers' mouths, making airtight seals.
6. Swirl all containers for 5 seconds to thoroughly mix the contents.
7. Every 5 minutes, examine the balloons. Write down and sketch what you see.
8. At the end of 30 minutes, stop the experiment. Summarize and analyze your results.
9. Make a list of all the factors that could influence the rate and amount of substances released during yeast fermentation. Based on the ingredients of each bottle, which bottle(s) would Martin Keller be most interested in? Explain. How might Martin Keller's research change the results of this experiment?

Journal Question During fermentation experiments, why can the balloon's inflation with carbon dioxide be used as a measure of ethanol production?

During fermentation, glucose and yeast produce carbon dioxide gas and ethanol.

$$\underset{\text{glucose}}{C_6H_{12}O_6} \xrightarrow{\text{yeast}} \underset{\text{ethanol}}{2C_2H_5OH} + \underset{\text{carbon dioxide}}{2CO_2}$$

Geothermal Energy

The word *geothermal* literally means "Earth heat." **Geothermal energy** is energy derived from the heat within Earth. Systems use the heat produced deep inside Earth to heat homes and produce electrical energy.

In locations where Earth's internal thermal energy gets close to the surface, underground water can be heated between 150–370°C (300–700°F) and placed under intense pressure. Wells are drilled at these locations to access this hot, pressurized water. When this water is brought to the surface, it quickly turns to steam. This steam is used to push turbine blades and spin a generator, like in a fossil fuel plant.

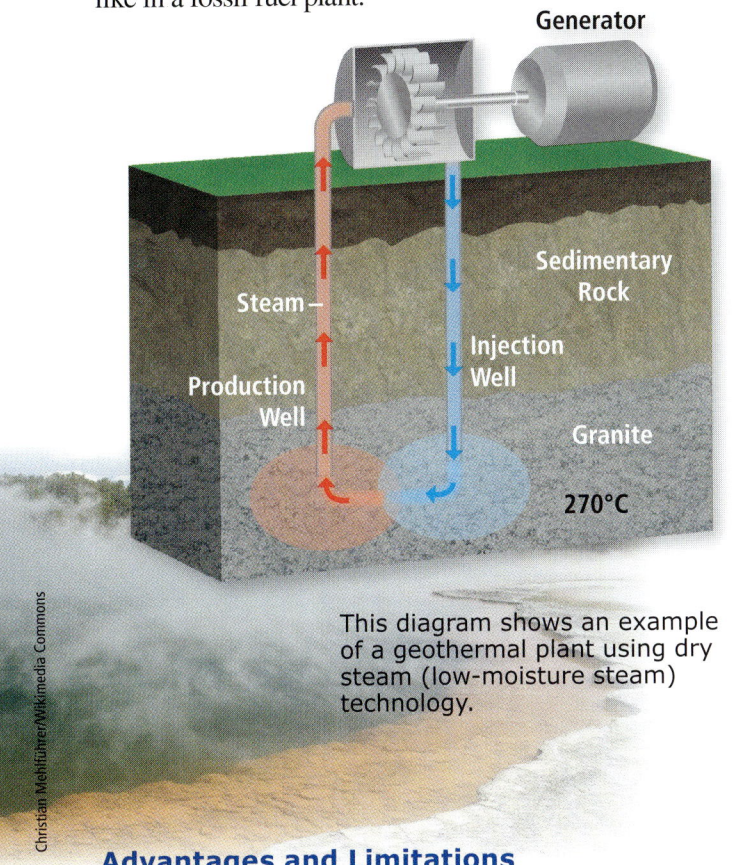

This diagram shows an example of a geothermal plant using dry steam (low-moisture steam) technology.

Advantages and Limitations

Geothermal power plants have many advantages. They do not burn fuel, and therefore minimize emission levels. Plus, the fuel does not need to be transported or piped to the power plant. Financially, once geothermal plants are built, there are no major additional costs.

One notable limitation of geothermal energy is that it is not readily available everywhere. However, with advancements in deep-drilling technology and geothermal detection, scientists and engineers are hoping to increase the number of areas that can begin using this energy source.

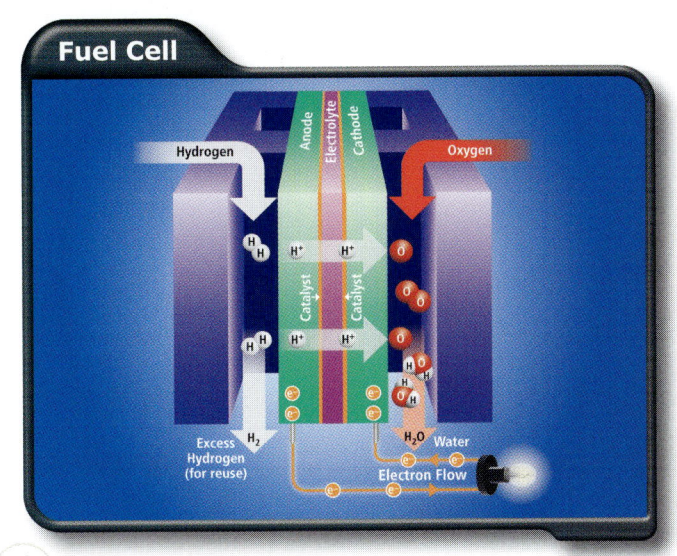

Fuel Cells

A hydrogen **fuel cell** is a device that generates electrical energy through the interaction between hydrogen and oxygen molecules. Both molecules are very different. Oxygen is a molecule that has a stronger attraction for electrons than hydrogen molecules. When combined in a fuel cell, a chemical reaction occurs in which hydrogen gives electrons to oxygen in a process that ultimately forms water. During this process, the electron flow from hydrogen to oxygen is channeled through an external circuit. The electrical energy generated from this reaction can be used to help power devices from cars to electronic systems on the International Space Station.

Advantages and Limitations

An advantage of hydrogen fuel cells is that no carbon dioxide or particle pollutants are released; only water and water vapor are emitted. Additionally, once hydrogen is available fuel cells are very efficient.

Currently, fuel cells are expensive, which is a big limitation when people are looking for cheap energy. Also, fossil fuels are presently the main source of hydrogen for fuel cells.

Making Decisions About Our Energy Portfolio

Our world is faced with many challenges and obstacles concerning energy. We are living in a world where more than 90 percent of our energy comes from sources that are non-renewable. Change must happen, but it must be well planned. Personal, political, economic, and environmental factors must be considered when analyzing, developing, and planning for energy changes. Scientists like Dr. Keller and his team of researchers are working hard to expand our energy options.

Mission 4: Energy Independence—The Quest for Sustainable Resources

Field Assignment

Enzymes Are Key

Recall that your mission is to *evaluate the future role of alternative energy resources*. Now that you have been fully briefed, it is time to survey your natural surroundings to see if the key for energy generation is sitting right there in your backyard.

Dr. Keller and the researchers at Oak Ridge National Laboratory (ORNL) are exploring ways to transform energy from renewable biomass into ethanol, a fuel gaining widespread popularity in transportation. In fermentation, simple sugars can be readily broken down to make ethanol. However, cellulose takes a long time to break down into simple sugars. Unfortunately, productive biomass plants in North America, like the poplar tree, generally store their sugars in the form of cellulose. Dr. Keller and his team are searching for microbes that possess the enzyme ideal for the rapid breakdown of cellulose into simple sugars. Their search has brought them to the hot springs of Yellowstone National Park.

Natural hot springs are often teeming with microbes that are unique in their ability to survive. Adapting to these extreme temperatures, they may have enzymes with the ability to break down cellulose into usable simple sugars faster, which is key for Dr. Keller's work.

The expedition team collected and isolated microbes from a decomposing poplar tree that had fallen into a hot spring. Back at the laboratory, the mystery microbe's genome was sequenced. As a result of the DNA sequencing process, Dr. Keller and his team found a very interesting enzyme that may provide information on the microbe's name as well as its function.

In this field assignment, you will use the protein information to help Dr. Keller identify the microbe found in Yellowstone National Park and assess whether it is suitable for further study in the breakdown of cellulose.

Objectives: To complete your mission, accomplish the following objectives:

- Describe, using secondary sources, the biological terms microbes, DNA, amino acid, protein, and enzyme and discuss how they relate to each other.
- Describe and analyze the amino acid sequence provided by Dr. Keller using online resources.
- Recommend prospective organisms for further research.
- Assess regions suitable for renewable and inexhaustible energy sources using Google Earth™.
- Create a presentation recommending the development and use of a renewable or inexhaustible energy source in a region of your choice.

Mission 4 Argonaut Field Assignment Video Join the National Argonauts as they explore the research facilities at Oak Ridge National Laboratory in Tennessee and meet Jaguar, the world's most powerful supercomputer dedicated to open scientific research, which can perform mind-boggling computations related to Dr. Keller's enzyme work.

Materials
- Mission 4 Field Assignment Data Sheet
- computer with Internet connection, Google Earth™, and presentation software installed

2. In order to identify the organism that this enzyme belongs to, scientists use a protein blast search, which compares amino acid sequences with a database of sequenced proteins.

   a. Go to the BLAST website. A link can be found in the JMC.

   b. Scroll down and click on the link that reads: "protein blast."

   c. Open a new tab and go to the JMC. There you will be able to copy the enzyme sequence.

   d. Paste the amino acid sequence into the field that reads: "Enter accession number, gi, or FASTA sequence."

   e. Click on: "BLAST" and wait a few moments while it searches for matches in the database.

3. Identify the top five organisms with the highest match. Click on each bar to find the organism's name.

Mission Challenge

Now that you have seen how Dr. Keller is looking to broaden the energy portfolio, you will use mapping software to look for more energy options.

1. Using Google Earth™, identify three to five regions in your country that have potential for renewable and/or inexhaustible sources of energy (search Google Earth™ for regions of high sunlight, geothermal activity, high winds, strong tidal flow, and consistent waves). Include justifications for your reasoning.

2. Investigate and describe how those three to five regions generate their energy today.

3. Propose an alternative source of energy for a chosen region assessing both benefits and limitations of using these alternative sources of energy.

Mission Debrief

Using all the information you have collected, create a presentation that recommends the development and use of alternative forms of energy in a region of your choice.

Journal Question What is the potential for renewable and/or inexhaustible forms of energy in your local area?

Amino Acid Code

Enzymes are a type of protein comprised of a long chain of amino acids. Scientists have created a code which uses letters to symbolize specific amino acids. There are 20 different types of amino acids. This is the enzyme that was found in the microbe Dr. Keller and his team of researchers brought back from Yellowstone National Park.

```
LNKLPRYKGFNLLGLFVPNMSYGF
FEDDFWMEEWGFNFARIPMNYRNW
YVEERPEIKEEVLEIIDKVVVWGQKY
GIHICLNIHGAPGYCVNEKTKEGYNL
WKDKEPLELFVSYWQTFAKRYKGIS
SKHLSFNLINEPRQYSKEEMTKEDFI
RVMTYTIEKIREIDKERLIIIDGVNYG
NEPVFELTSLGVAQSCRAYLPFELTH
YKAEWVEGSDKFSEPSWPLVRDNGE
VIDREYLRRHYEKWTKLFDYGVGVI
CGEGGAYKYTSHEVVLRWLSDVLD
VLKELNIGIALWNLRGPFGIIDSGRE
DVEYEDFYGHKLDRKLLELLMRF
```

Field Preparation

1. Using the enzyme that Dr. Keller sequenced, identify the total number of amino acids that make up this enzyme.

Connections

Renewable Resources

Harnessing the Power of Plants

Many types of renewable and inexhaustible energy resources exist. Examples include solar, wind, waves, geothermal and biomass. Cellulosic biomass is the most abundant biological material on Earth, and can be used to make biofuels. There are two main types of biofuels, ethanol and diesel. These can be used as a substitute for non-renewable energy resources such as gasoline and diesel fuel made from petroleum.

An advantage of biofuels is that they are made from plants. As plants grow, they absorb carbon dioxide (CO_2) from the atmosphere. This means the biofuel they produce is virtually "CO_2 neutral."

Mike Goosey, an innovation manager at Shell, explains more about biofuels – how they are produced and what developments are being made in the next generation of biofuels.

"Today's most widespread biofuel, ethanol, is commonly made from sugar cane, corn or wheat. These plants are converted into ethanol via a process called fermentation. The second most widespread biofuel is biodiesel which is made from the vegetable oils in plants such as palm oil or soybeans.

"In theory, ethanol and biodiesel can be used in pure form for transport fuel. However, the reality is that most of today's engines are not designed to use them. Also, they have some shortcomings as a fuel: for example, ethanol absorbs water, which can cause engine corrosion, and biodiesel has a tendency to freeze. However, ethanol can be blended at low concentrations with gasoline and biodiesel with diesel – and standard engines can accommodate a blend of 5%-10% with no problem."

88 • Operation: Infinite Potential www.jason.org

Next Generation

Today, many scientists are working to develop the next generation of advanced biofuels. They are trying to make biofuels from sustainably grown crops, such as switchgrass and poplar trees and even algae. These crops are leading choices for making biofuels, as they do not compete with the farming of food crops.

Mike explains,

"In these plants, there are two basic parts that can be used to make biofuels: the cell walls of woody plants and the vegetable oils found in algae or other inedible, oily plants such as jatropha."

Scientists are researching how to improve the production of biofuels by investigating improving crop yields and trying to reduce the amount of resources or energy put into growing the crop. For example, by reducing the amount of fertilizer or water needed to generate the biomass, resources would be saved.

Next Challenge

According to Mike, the next challenge is researching better ways to make biofuels. Scientists are looking at two ways to tackle this: one using biology and the other using chemistry.

"An example of the latter is something that we are developing with another company. We gasify biomass, typically wood chips, to produce syngas."

When biomass is gasified, it reacts with oxygen and steam. The biomass chemically changes to produce a mixture of carbon monoxide and hydrogen, called synthesis gas or syngas. The syngas is then processed further, where it reacts and grows into long chains of up to 200 carbon atoms. These chains form the basis of the biofuel.

In the biological process to make ethanol, enzymes break down cellulose into simple sugars. These sugars can then be fermented by yeast into ethanol. Researchers are looking to other industries, like the pharmaceutical industry, for processes and techniques that may help with the next generation of biofuels production.

A biofuel-powered prototype and winner of the Shell Eco-Marathon with an energy consumption equivalent to 2,885 km/l of petrol.

Your Turn

Alternate energy sources require the investment of money, in addition to the creativity and ingenuity of the scientists and engineers to become part of the energy portfolio. Select an alternate energy source you would advocate for and investigate how you could fund its development.

Connections: Renewable Resources • 89

Mission 5
Energy Security
Powering Our Future

"Your energy consumption is affected by every aspect of the things we make and buy – from materials to design to your choices as a buyer."

—Constance Adams
National Geographic Emerging Explorer
Space Architect

Constance Adams

Constance Adams, a NASA consultant and National Geographic Emerging Explorer, is a space architect who designs space modules and off-world habitats that are energy efficient and functional for the space crews that inhabit them.

Meet the Researchers Video
Constance Adams will help you discover how your choices and the choices of others can impact local and global energy resources.

National Geographic Emerging Explorer

Read more about Constance online in the JASON Mission Center.

Photos above (left to right): NASA; Elijah van der Giessen/Wikimedia Commons; Peter Haydock, The JASON Project; NASA; Albert Moldvay, NGS; Wikimedia Commons; Mark Thiessen, NGS; David Boyer, NGS

Your Mission...

Create a blueprint for a secure energy future.

To accomplish your mission successfully, you will need to

- Evaluate the current and future state of energy supply and demand.
- Examine strategies that will ensure sufficient and affordable energy.
- Design and model systems to better understand energy transfers and transformations.
- Discover the geopolitics, environmental factors, and economics that shape energy policy.
- Explore energy efficiency and understand its role in conservation.
- Measure the impact of behavioral changes on energy demand.

Mystery Connection

ALERT!

You are the captain of an interplanetary mission. Halfway to Mars, there is an explosion in the spacecraft. For an instant, the power goes out. It returns, but the interior lighting has dimmed and many of the systems have shut down.

The computer indicates that electricity generation has been compromised. Possibly half of the fuel cells that meet the craft's energy needs have shut down.

The ship carries two craft within its storage module—a Martian lander and an orbiting communication satellite. The Martian lander contains a separate life support and electrical supply system. The communication satellite has an extensive solar array.

Based upon the available resources, develop a plan to conserve energy that would ensure the safety of your ten-person crew. How could you conserve energy? What sort of backup energy sources might you tap? How would you generate enough energy to make your return trip to Earth?

Mission 5: Energy Security—Powering Our Future • 91

Powering Our Future

"Three...two...one...blast-off!" As the engines roar to life, a rocket lifts from the launch pad. With a slow and steady acceleration, it gains altitude. Higher and higher it climbs until it can no longer be seen with the unaided eye. Eight minutes after lift off, the engines shut down and the craft silently enters Earth's orbit.

Circling their home planet, the crew readies the spaceship for its extended mission. A command is entered into the computer. In response, the craft begins a radical transformation. Like a life vest filling with air, the ship expands in size. When the fabric shell reaches its final size, the living quarters have increased more than threefold! The craft is a TransHab Module—an inflatable space structure that expands to create additional living space. With this expansion, there is enough room to house a crew of six for a journey to Mars.

Back on Earth, resources are sufficient for all the planet's inhabitants. Advances in technology, along with effective conservation policies, have improved the quality of life. We have met the needs of a global population through advances in energy generation, distribution, and use.

Envisioning this type of future and creating the innovations that will make it happen are all in a day's work for Constance Adams. That is because Constance is a space architect who works with NASA. Her out-of-this-world projects focus on designing innovative habitats used for extended space flights. Her designs for Earth structures may help conserve energy, ensuring that valuable resources will not be wasted. To create these novel solutions, Constance integrates disciplines including engineering, architecture, and the social sciences. By drawing on these different fields of study, she creates structural designs that best address the world's needs and available resources.

Mission 5 Briefing Video Prepare for your mission by viewing this briefing on your objectives. Learn how explorers like Constance Adams use their understanding of energy to make choices about energy-efficient designs and the energy consumption involved in using products and services.

Join the Team
At NASA's Johnson Space Center in Houston, Texas; Hannah Zierden, Cindy Parish, Constance Adams, Anthony Lopez, and Madhu Ramankutty (L to R) discuss the factors that Constance must consider when designing space habitats that need to remain cool while constantly being bombarded by the sun's electromagnetic energy. The structures that Constance designs both utilize and conserve valuable energy resources.

Peter Haydock, The JASON Project

Mission Briefing
Energy Supply and Demand

Constance Adams understands that every decision made has an impact. As she designs for off-planet missions or buildings on Earth, she has to consider how best to supply the right amount of energy to her designs. She considers weight-saving materials, low power lighting, using renewable and inexhaustible energy resources, and using passive solar heating to maintain the right temperature for the astronauts. She even looks to use locally available materials to avoid transportation costs. Every decision she makes not only impacts the mission but anyone that may also use those resources. Similarly, as we consider our energy portfolio, we will need to understand what impact our decisions have in our homes, in our community, and across the globe.

92 • Operation: Infinite Potential   www.jason.org

Energy Demand

A nation's economy is closely tied to its energy consumption or demand. In general, the larger an economy is, the more energy is consumed per person. Today, the United States is the world's largest consumer of energy. Each year, the United States consumes about 100 quadrillion Btu of energy, or **quads** for short. That is more than one-fifth of the total energy consumption for the world.

On average, each United States citizen uses more than 300 million Btu annually. It sounds like a lot—and it is—especially if you compare it to the energy consumed by a citizen of Haiti, who uses about one-hundredth that amount. These energy demands come from home heating, transportation, health and nutrition, electricity, and the manufacture and supply of goods.

Energy Supply

Throughout the world, energy demand depends on a variety of limited resources. Most of our energy demand is supplied by oil, natural gas, and coal sources. The remaining is supplied by nuclear, renewable, and inexhaustible sources.

However, oil, gas, and coal are limited resources that are scattered across the world. The uneven distribution of—and world-wide demand for—these resources have made them commodities that can be traded between countries.

In 2006, the world's top oil producer was Russia, which extracted and produced about ten million barrels per day. About half of this oil was exported to other countries. Saudi Arabia extracted and produced just over nine million barrels per day and exported almost all of it—approximately eight million barrels per day.

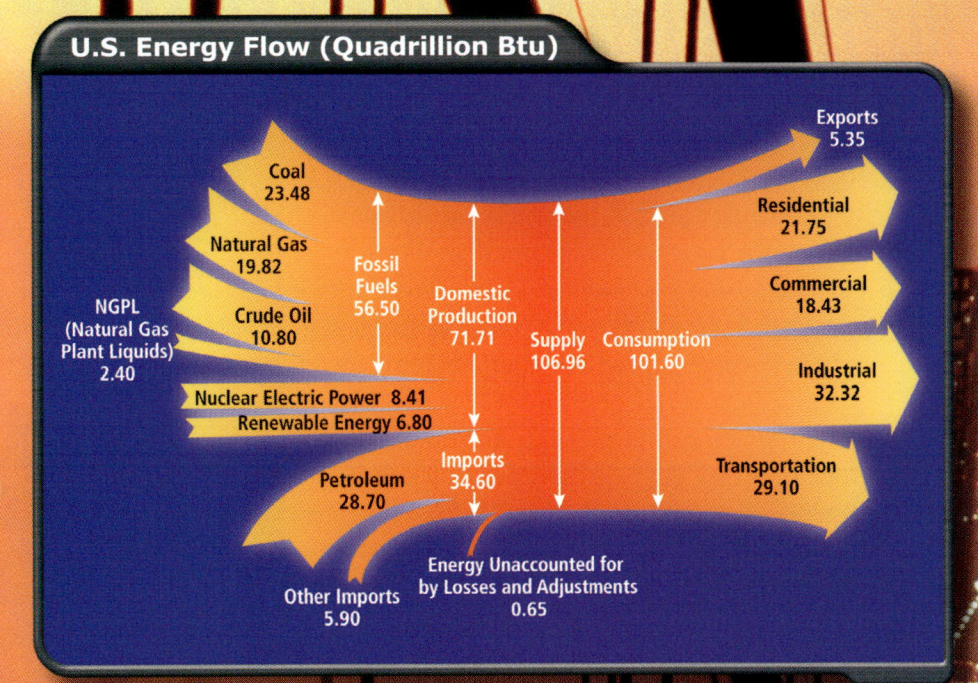

U.S. Energy Flow (Quadrillion Btu)

- Coal 23.48
- Natural Gas 19.82
- NGPL (Natural Gas Plant Liquids) 2.40
- Crude Oil 10.80
- Fossil Fuels 56.50
- Nuclear Electric Power 8.41
- Renewable Energy 6.80
- Petroleum 28.70
- Imports 34.60
- Other Imports 5.90
- Domestic Production 71.71
- Supply 106.96
- Consumption 101.60
- Energy Unaccounted for by Losses and Adjustments 0.65
- Exports 5.35
- Residential 21.75
- Commercial 18.43
- Industrial 32.32
- Transportation 29.10

World Primary Energy Consumption

Region	Annual Energy Consumed	Population	Annual Energy Consumption Per Capita
North America	121 quads	442 million	276 million Btu
Central and South America	24 quads	460 million	53 million Btu
Europe	86 quads	593 million	146 million Btu
Eurasia	46 quads	285 million	160 million Btu
Middle East	24 quads	191 million	127 million Btu
Africa	14 quads	934 million	16 million Btu
Asia and Oceania	156 quads	3,691 million	43 million Btu

Source: 2006 EIA Consumption Data; 2007 EIA Population Estimates

Mission 5: Energy Security—Powering Our Future

Energy Dependence

This supply and demand of international goods has connected economies around the world for decades. International trade for energy resources has become increasingly important for countries, like Japan, that do not have large oil reserves but use a lot of it. Like the United States, they depend on other countries for oil.

The United States consumes over 20 million barrels of crude oil every day. Approximately one-third of this amount comes from domestic oil fields located mostly in the Gulf of Mexico, Texas, and Alaska. Because domestic oil extraction and production does not meet all of this demand, they must rely on other countries. Canada provides the United States with one tenth of its oil and most of the remainder comes from Saudi Arabia, Mexico, and Venezuela.

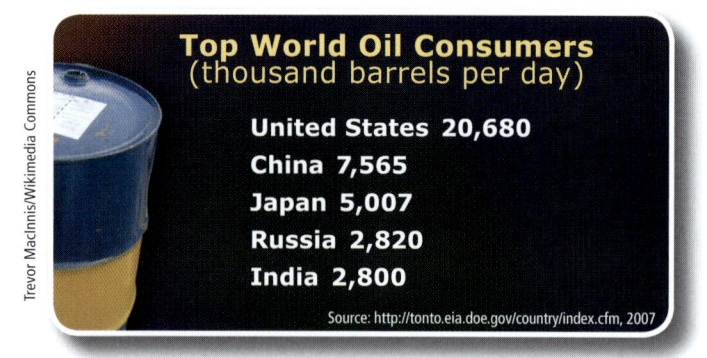

Top World Oil Consumers
(thousand barrels per day)

United States 20,680
China 7,565
Japan 5,007
Russia 2,820
India 2,800

Source: http://tonto.eia.doe.gov/country/index.cfm, 2007

Trevor MacInnis/Wikimedia Commons

Energy Security

It is critical that a nation manages these resources wisely. Access to a reliable source of energy is called **energy security**. This supply must not only meet demand, but also be affordable. If the access to energy resources is too small, inconsistent, or expensive, then energy security is jeopardized. Energy security can be further compromised by outside influences, including wars, political conflicts, and even natural disasters. If this happens, there could be drastic effects, like crippling the economy and lowering the standard of living.

Diversifying the Portfolio

Although the United States is today's top consumer of energy, things are changing. Increasing populations and advancing economies, especially those of India and China, will greatly affect the production and consumption of energy resources. It is estimated that by 2030, China and India alone will account for

Fast Fact

Do you have something set aside for a rainy day? The United States government does. The Strategic Petroleum Reserve (SPR) is an emergency supply of oil managed by the United States Department of Energy. In September 2005, Hurricane Katrina devastated New Orleans, Louisiana and the surrounding Gulf Coast. In addition to its impact on life and property, this monster storm shut down the region's oil supply, production, and refining facilities. To meet the nation's ongoing demand for fuel, the SPR was tapped and its stock made available to the oil industry. Other countries around the world also have strategic oil reserves or are developing strategic reserve plans for times of need.

a 45 percent increase in global energy consumption. The predicted increase of coal use in China and oil use in India will affect world markets by reducing the supply for other countries and accelerating the need to move toward alternative energy resources.

Based on all the issues of energy supply, demand, dependence, and security, it is apparent that a diverse energy portfolio is important. Energy sources, such as wind, solar, geothermal, biofuels, hydroelectric, and fuel cells, are all becoming attractive alternatives to fossil fuels. Currently, renewable resources provide around seven percent of the world's present energy demand. Further increasing and diversifying the use of these resources as energy sources will increase energy efficiency to maximize and manage existing resources.

Lab 1

Cooling Off

Depending on which side is facing the sun, the exterior temperature of a spacecraft can vary by hundreds of degrees. This can affect the interior temperature of the spacecraft as well. Constance Adams uses her knowledge of energy transfer and transformation to help her design ways to best regulate the internal temperature and maintain livable conditions.

To maintain a life-supporting temperature, Constance considers a design that uses fluid to absorb, transport, and release thermal energy. If she can use a natural exchange of thermal energy, she can conserve energy resources which would have been spent on heating and cooling the craft's interior.

In this activity, you will explore natural heating by the sun and investigate how a fluid-cooled system works.

Materials
- Lab 1 Data Sheet
- beaker
- black paper
- scissors
- duct tape
- squeeze bottle
- bowl
- insulated container (such as a cooler)
- ice water
- plastic wrap
- plastic tubing
- 2 thermometers
- box

Lab Prep

1. Obtain the prepared box from your instructor. Insert tubing through the box, so that it extends from both holes. Use duct tape to seal around the area where the tubing enters and leaves the box.

2. Place a thermometer in the box. Stretch plastic wrap over the box top, forming an air-tight enclosure. Secure the plastic to the outside box sides with duct tape.

3. Fill the insulated container with ice water.

Make Observations

1. Take all materials outdoors. Measure the initial temperature inside the box as soon as you expose the box to sunlight.

2. Monitor and record the temperature inside the box every five minutes until the temperature stops changing. Once the temperature stops changing, use the second thermometer to measure the temperature of the ice water solution.

3. Fill the squeeze bottle with ice water. Make a prediction regarding how filling the plastic tubing with this ice water will affect the temperature inside the box.

4. Upon your instructor's signal, inject this ice water into the tube, stopping when water flows out the other side of the tubing. With the ice water in the tube, carefully observe and record the temperature inside the box at one-minute intervals. Stop recording the temperature in the box when the temperature starts to climb.

5. Remove the water from the tube and collect it in the bowl. Measure the volume of this water and record its temperature.

6. What variables can be changed to improve the efficiency of this setup? Discuss your ideas with your instructor, and then proceed with your investigation.

Journal Question How might the concepts observed in this lab be applied to space heating and cooling?

Mission 5: Energy Security—Powering Our Future • 95

Energy Efficiency

Another way to meet energy needs is to cut back on demand. You can do this by buying "green." Suppose you are shopping. You are in an electronics store, helping to pick out a new TV. Which one should you choose? These days, more people are considering energy efficiency as part of their purchasing decision. Although you might think that energy efficiency does not really matter when it comes to TV sets, it does. Powering up a large plasma screen for only a few hours can use more electricity than a full-sized refrigerator consumes in a day! Over a year this can really add up.

Like other physical qualities, energy efficiency can be measured and expressed as a mathematical relationship. When it comes to an appliance like a refrigerator, the system must complete a certain amount of work. The ratio of work output to energy consumed per unit time is a measure of its efficiency. For example, if a refrigerator requires 400 W to cool its contents, it is twice as energy efficient as a unit that requires 800 W to do the same work in the same amount of time.

ENERGY STAR®

So how do you decide which TV set is most efficient? One way is to look for the ENERGY STAR® label. The ENERGY STAR® label is awarded to energy efficient devices and structures. Designs which have earned this label are more efficient, using less energy, and ultimately save the consumer money as well as help protect the environment.

Fuel Economy

On today's new cars, you will see an Environmental Protection Agency (EPA) window sticker that displays fuel economy estimates. Consumers can use this information to compare the energy efficiency of different vehicles and estimate operating costs. For cars sold in 2009 in the United States, fuel efficiency varied from 8 mpg (miles per gallon) for a sports car, to 48 mpg for a compact hybrid. A car that gets more miles per gallon ultimately uses less fuel and is more efficient.

Hybrid vehicles are powered directly or indirectly by two or more energy sources. The combustion engine burns gasoline like a standard vehicle. The subsequent energy released spins the wheels and charges batteries, which also operate the car. Onboard computers continually monitor the charge level of the batteries, as well as driving demands. Based upon this data, a computer switches the source of energy between gas and the electric batteries to maximize fuel efficiency. This means that the car does not always run on gas.

Making Decisions

Did you forget to turn off a lamp or appliance last night? Most of us often overlook shutting down one or more energy-consuming devices that are not in use. When you leave an appliance on longer than necessary, you are not just wasting electrical energy, you are paying higher electric bills and increasing your carbon footprint.

Team Highlight
Argonaut Madhu Ramankutty examines the design qualities of an astronaut's space suit.

How do you make decisions? Do you let the flip of a coin decide the outcome? You probably do not. More likely, you follow a thinking process that helps you arrive at a valid and fair-minded decision. You can apply that process to an energy decision.

Compact Fluorescent Lamps

Most likely, you have seen a **compact fluorescent lamp** (CFL). These devices are energy-saving alternatives to standard incandescent lamps. Although CFLs cost more up front than incandescent lamps, they tend to last up to 15 times longer while producing the same amount of light at a lower wattage. This means that over the life of the CFL, it will probably cost your family less to replace the CFL than to purchase several incandescent bulbs. As a bonus, it even uses less electricity.

How could you find out how much you would save by investing in a single CFL? You might begin by comparing the operating costs of a 150 W incandescent bulb to the CFL equivalent 30 W. Using the figures on this page, you can see that the 30 W CFL will save you $23.66 in yearly electrical costs. As far as your electric bill goes, it seems like a wise decision to upgrade your lights. What other factors need to be considered? Will the cost of the newer bulbs offset any gains in energy efficiency?

Math Connection

To determine the amount of energy consumed daily, use this formula:

wattage X hours of daily use X $\frac{1 \text{ kW}}{1,000 \text{W}}$ = daily kilowatt-hours (kW•h) of power consumption

Incandescent $\frac{(150 \text{ W})(6 \text{ hours})}{1,000}$ = 0.9 kW•h

CFL $\frac{(30 \text{ W})(6 \text{ hours})}{1,000}$ = 0.18 kW•h

To determine the yearly cost, multiply the daily kW•h usage by the number of days in a year and the price paid for a kW•h, say 9 cents.

(0.9 kW•h) (365 days) ($0.09/kW•h) = $29.57

(0.18 kW•h) (365 days) ($0.09/kW•h) = $5.91

Savings per year = $23.66

Light Emitting Diode

Light-emitting diode (LED) based lights are another form of technology on the market today. You may have seen these already in stoplights and the brake lights of trucks. This technology has been around since the 1960s, but with recent advancements, the lamps have become much brighter. Home and business lighting solutions appear to have even better energy savings with LED than with CFL technology.

As you can see, making a good decision depends upon understanding and weighing options. Only then can you make an informed and fair-minded decision. Remember that committing to an option is not the end of the process. Good decision-makers continue to review their choices, evaluating both successes and failures.

TransHab and Its Evolution into Genesis I

In order to withstand and survive the harsh environment and challenges associated with long-term space travel, a team of NASA scientists, led by space architect Constance Adams, designed the TransHab. The TransHab combines concepts and ideas from a variety of disciplines in order to meet the needs of humans traveling through space for extended periods of time. Constance and her team addressed architectural, engineering, industrial, and sociological factors in order to create the blueprint for a safe, comfortable, and self-sufficient living quarters for astronauts.

In 2001, NASA collaborated with Bigelow Aerospace to improve on the TransHab design. The TransHab was later renamed Genesis I and on July 12, 2006, Constance Adams' remarkable vision was successfully launched into space, and is currently orbiting Earth.

Launch date: July 12, 2006

Launch location: ISC Kosmotras Space and Missile Complex, Russia

NORAD identifier #: 29525

Speed: 27,243 km/h

Orbit: 563 km above Earth's surface

Dimensions contracted: Length 4.4 m, Diameter 1.6 m

Dimensions expanded: Length 4.4 m, Diameter 2.54m

Usable volume: 11.5 cubic meters

Thickness of shell skin: 15.2 cm

Shell skin material: Multiple layers of Kevlar (designed to be airtight and tough—able to withstand space debris including small meteorites)

John Frassanito and Associates

Mission 5: Energy Security—Powering Our Future • 97

Heating Spaces

How is your home heated? Do you depend upon natural gas as a heating fuel? If not, perhaps you use oil, propane, or electricity. Some people even depend entirely upon the warmth of wood stoves. No matter which energy source you use, there is a cost to operate your heater, and sometimes this cost can really add up. For most households, heating is the largest energy expense.

Example

Aware of **heating costs and efficiency**, Constance Adams incorporates a variety of energy-saving strategies into her architectural designs. She uses renewable and inexhaustible sources of energy, such as sunlight, to help meet heating demands. She also considers the role of natural surroundings such as earth materials, to help insulate structures. Additionally, she uses passive solar heating to reduce energy demand.

Natural Gas

More than half the homes in the United States meet their space-heating needs with natural gas. Most of these homes have a central warm-air furnace. Natural gas burned within this device produces thermal energy and transfers it to the air. The warm air is then circulated through the home using fans and vents.

Team Highlight

Inside a model of one of the International Space Station (ISS) modules, Argonauts Hannah Zierden, Anthony Lopez, Cindy Parish, and Madhu Ramankutty (L to R) explore how astronauts live in the extreme environment of space.

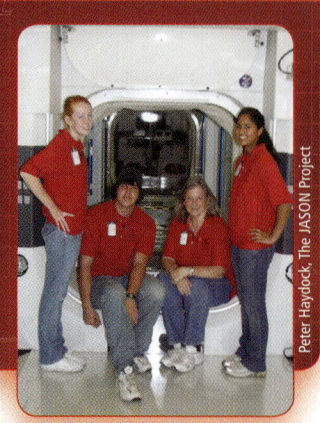

Electrical Heat

Electrical heat is the next most common heating method, meeting almost one-third of the United States space-heating demands. Unlike the heat produced by the burning of fossil fuels, using electrical resistance to transform electrical energy into thermal energy is a process that is nearly 100 percent efficient. However, the steps that generate and transmit electricity to your home are inefficient, resulting in energy losses as high as 70 percent. So by the time it is transformed into thermal energy, electrical heating can be more expensive than the on-site burning of fossil fuels.

International Connection

LOCATION Around the World

Often touted as being "as clean as water" and the "fuel of the future," high hopes are held for hydrogen power—from running your car to heating your home. While hydrogen does hold promise, the technology has a way to go before it becomes a viable fuel.

Hydrogen first needs to be prepared. The most common way of producing hydrogen is from natural gas. Steam is used to convert hydrocarbons into hydrogen and carbon. Around half of the world's hydrogen is produced in this way. New hydrogen can also be produced from water, where it is extracted through electrolysis.

Hydrogen then can be used in hydrogen fuel cells. It is not yet practical to use fuel cells for home energy conversion, but they can be used in cars. In hydrogen fuel cell vehicles, a chemical reaction inside the fuel cell—usually between hydrogen and oxygen—creates electricity for the motor, and the only resulting tailpipe emission from the vehicle is water vapor.

Shell, a pioneer in hydrogen development, expects that, one day, hydrogen will be produced from wind, solar or even biomass.

Operation: Infinite Potential   www.jason.org

Lab 2

Making Models

Scientists like Constance Adams use many different types of models in their research and work. Mental models are used to envision concepts, relationships, and structures. Sketches are used to record these concepts on paper and communicate ideas to fellow scientists and designers. Diagrams, such as architect blueprints and Computer Aided Design (CAD) files help Constance with scale measurements of real world objects—either built or imagined. In her work at NASA, Constance uses all of these model types to develop and communicate her ideas of terrestrial and space-based architecture.

In this activity, you will make a variety of models to design a module for the International Space Station (ISS). These models will help you think about, create, and communicate your ideas and concepts to your fellow scientists and designers.

Team Highlight
Anthony Lopez navigates one of the model tools found on board the International Space Station.

Materials
- Lab 2 Data Sheet
- ruler
- materials provided by instructor

Lab Prep

1. Use print and online resources to learn about the history, role, modules, and future of the ISS.

2. Discuss with your classmates how solar panels are used by the station to obtain sufficient energy to meet its operational and life support systems.

3. Research and write about the challenge of energy generation, transfer, and transformation in space.

Make Observations

1. Examine a photo of the ISS. Based upon this image, create a diagram that illustrates the craft. Label and identify the parts and structures of the station.

2. From your research, create a list of the current modules that comprise the station. Describe the role of each.

3. Select one of your listed modules. Based upon your research and your understanding of its role and its overall appearance, draw a set of blueprints that could illustrate the interior of this module.

4. On your blueprints, identify the flow of all energy transfers and transformations within the module.

5. Exchange blueprints with another group. Discuss the strengths and limitations of each other's design.

6. Update your set of blueprints as needed. Then, using a variety of materials, assemble a 3D representation of your module.

7. Critically analyze your scale model. Can a model like this be used in an investigation to learn more about the full-sized habitat? Explain. What sort of modifications would you need to make in order to transform this model into a subject for experimental testing and inquiry?

Journal Question What can we learn from the International Space Station about meeting our planet's energy needs?

Mission 5: Energy Security—Powering Our Future

Our Energy Future

As scientists work to expand our energy portfolio options, others are developing advanced distribution and storage technologies. These efforts will help make energy distribution more efficient, and will also help make energy supply more consistent.

A Smart Electric Grid

To reach a sustainable energy future, engineers are exploring a more effective method of distributing electrical energy. They envision a **smart electric grid,** a digital electricity distribution and management system that would connect all sources of energy generation to all types of end users.

The electrical energy pouring into the grid would not be limited to the energy generated by large municipal power stations as it is now. Instead, electrical generation capabilities will be spread across the nation. Individual homes capable of producing electrical energy through technologies, such as solar panels and wind turbines, will be able to contribute to this grid, supplementing the power produced by larger stations. Perhaps your school roof will be painted with a thin solar film that transforms sunlight directly into electrical energy. Maybe a small wind turbine will occupy a corner of the schoolyard. Whatever technology is used, it will be tied into a national or global grid system, allowing everyone to share energy resources.

Energy Storage

As more electrical energy generation shifts to renewable sources, the consistency of its generation will become an issue. For example, when the air is still, wind turbines are motionless. To compensate for inconsistent energy generation, improved energy storage technologies will be needed, which could be seamlessly tapped to meet energy demand.

In addition to a variety of chemical storage strategies, there are several methods of mechanical storage. Excess energy can be used to pump water to a higher elevation. Then, when energy is needed, the water is released. As it flows downhill, the water spins a turbine which powers a generator. Another mechanical solution depends upon compressed air. Surplus electrical energy can be used to energize pumps that compress air. When electrical energy is needed, the pressurized air is released and heated. The expansion of the gas spins turbine blades, resulting in electricity production.

> **Example**
>
> You are probably familiar with a battery as an **energy storage** device. Although it might meet household needs, the power grid in the future will require a larger and more efficient means of energy storage. Fuel cells could be that storage solution. During peak energy production times, excess energy could be used in a reaction that splits water into its component elements—hydrogen and oxygen. Later on, when energy is needed, this reaction could be reversed, combining the hydrogen and oxygen molecules to form water. This reverse reaction would generate electrical energy that could be transmitted through the power grid.

Lab 3

Using the Sun's Power at Night

Currently, spacecraft and modules only have two choices for energy production to support the systems and crew—solar and nuclear. Much like on Earth, Constance Adams must assess the advantages and limitations of these energy sources when designing new space modules. For modules close enough to exploit the sun's energy, such as the International Space Station, the sun provides a free, non-polluting, inexhaustible energy source. With her forward thinking, Constance strives to design new habitats for future space modules that can best exploit this supply of energy.

In this activity, you will have the opportunity to observe and analyze the components of a solar lawn light. After becoming familiar with these devices, you will work to investigate factors that influence and affect their operation during the day and night.

Materials
- Lab 3 Data Sheet
- solar lawn light (with removable battery)
- 2 wire connectors (with alligator clips)
- multimeter
- sheet of opaque paper

Lab Prep

1. Examine the lawn light supplied by your instructor. Is this the entire lawn light, or only one part of the complete structure? Explain.

2. Make two diagrams of this assembly. One diagram should illustrate the components that are located on the top surface. The other diagram should illustrate the components visible on the lower surface.

3. From what you observe, can you infer the inner wiring that is blocked from view? Make another diagram that illustrates how the parts of the solar lawn light might be wired together.

Make Observations

1. Place the lawn light upside down on the table top so that the photovoltaic (PV) cells are not exposed to light.

2. Attach the battery compartment terminals to a multimeter. Set the multimeter to measure voltage. Describe and explain your observations.

3. Turn the lawn light over, exposing the PV cells. What happens now? How many volts are generated by the light in the room? Describe and explain your observations.

4. Cover half of the exposed cells with a sheet of opaque paper. How was the generated voltage affected? Explore the relationship between exposed cells and generated voltage.

5. Use the multimeter to determine the optimal location in your surroundings for generating voltage. Is there another way of further increasing the amount of generated voltage? Explain.

6. Tilt the assembly so you can see the light-emitting diode (LED). Cover the light sensor while leaving the PV cells exposed to light. Does the LED light up? Explain.

7. Remove the connections and flip the assembly so that no light reaches the PV cells.

8. Insert the fully discharged battery into its compartment. Does the lamp illuminate now? Explain.

9. Develop a strategy to determine the relationship between the length of time the battery is charged and the length of time the LED remains lit.

10. With your instructor's approval, proceed with the inquiry.

Journal Question How can a solar lawn light be used to model the strengths and limitations associated with alternative energy resources?

Mission 5: Energy Security—Powering Our Future • 101

Argos to the Rescue

Calling all Argos! It is time for action. There are many ways that you can help conserve energy resources. Although the actions may seem small, they do add up. With the combined efforts of Argos around the world, we can make a huge impact! Here are some of the ways that you can reduce your energy consumption.

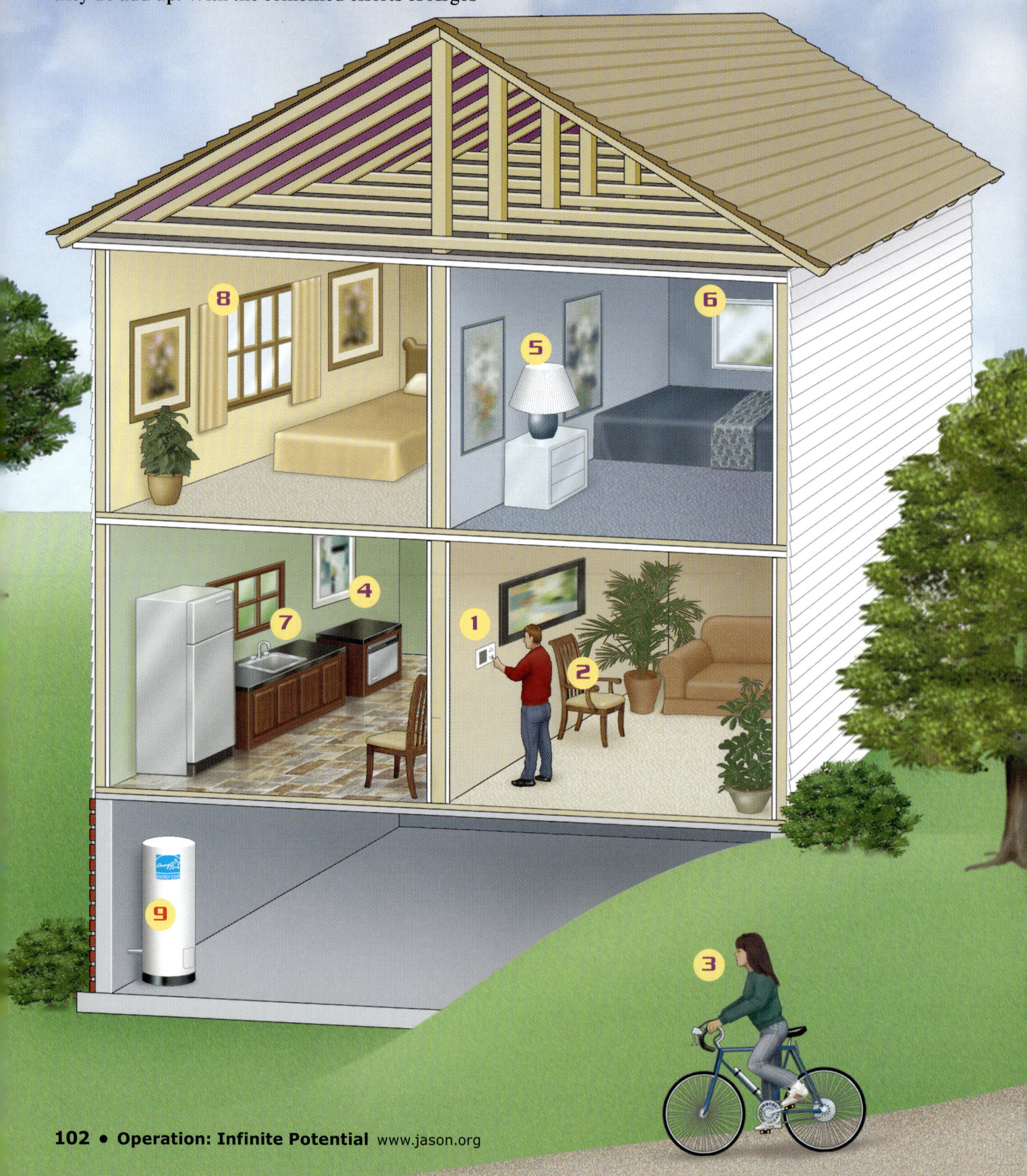

102 • Operation: Infinite Potential   www.jason.org

What You Can Do

1. **Reset your home thermostat.** During winter, set your home heating system for a warm and comfortable home—not a hot home. Adjust your home heating thermostat to 20ºC (68ºF) for the hours that you are awake. At night, you can lower the thermostat a few degrees. During the summer, set the thermostat to 26ºC (80ºF) and use low velocity fans for a cool home—not a cold home. At night, use a fan to keep you cool instead of air conditioning. If you have a programmable thermostat, you can set it to automatically change settings at different times of the day.

2. **Wear the right clothing.** During cooler months, it is important to stay warm both outside and inside your home. Indoors, wearing additional layers will help you stay warm so you can set the thermostat even lower. At night, use an extra blanket and warm pajamas to better trap body heat.

3. **Use "green" transportation.** When you can, walk or ride your bike. If riding in a vehicle is unavoidable, consider carpooling or public transportation options.

4. **Conserve electricity.** Do not forget to turn off lights and other electrical devices when not in use. The computer that remains on overnight continually uses electricity. If you use a dishwasher, only run it with a full load. When the dishes are done, open the door and let the dishes air dry instead of using the heated dry feature. Likewise, only run full loads in the clothes washer. Setting the water in these appliances to warm instead of hot saves additional energy. Check out the appliance controls, and whenever available, use energy-saving settings.

5. **Replace incandescent bulbs with compact fluorescent lamps (CFLs) and light-emitting diodes (LEDs).** Get adults involved and discuss which standard light bulbs can be replaced with more efficient CFLs or LEDs.

6. **Fix air leaks and drafts around the house.** In winter, cold drafts can quickly chill a warmed home. Likewise, warm air leaks during the summer can counter the effects of an air conditioner. Use self-sticking foam or vinyl weather stripping to seal off any leaks around windows and doors.

7. **Stop hot water energy losses.** Do any of your faucets leak? They could be leaking hot water. In most cases, replacing a worn washer fixes the problem. With an adult, inspect your house's hot water heating system. Are the water heater, hot water storage tank, and hot water pipes covered by insulation? If not, you could be losing a large amount of heat from the system.

8. **Draw the curtains.** During the winter, open curtains and window shades to allow light into your house. This solar energy will naturally heat the indoors. When the sun goes down, draw the curtains and shades to help retain this thermal energy.

9. **Be a "green" consumer.** Next time something needs to be replaced, think about your decision. Is it something that can be fixed? If it must be replaced, consider energy-efficient options. If a product or its packaging needs to be discarded, can it be recycled?

10. **Buy locally.** Transporting goods from far away expends greater energy and creates pollution.

11. **Share your understanding.** You have an opinion, and your opinion counts! Discuss with adults the different ways we can be more efficient with our energy consumption. Help them make informed and fair-minded decisions.

Landscaping for Energy Savings

Landscaping is not just for aesthetics. It can also assume a role in a building's energy efficiency. Planting trees which create shade can help a structure naturally stay cool. Plus, a row of trees can create a wind block, which reduces energy losses during cooler months.

Lab 4

Communicating with Graphics

Creating an energy secure environment within a deep space module is a concern for Constance Adams. Working with engineers, scientists, and fellow architects, Constance studies a module's energy budget detailing both energy production and consumption. Constance uses and creates diagrams, illustrations, and charts to share her thoughts and data with other architects and scientists.

In this activity, you will have the opportunity to examine a chart that communicates energy supply and demand in the United States. From this chart, you will learn more about the United States' energy sources and sectors. You will import this data into spreadsheet software, and use it to create additional graphics to analyze our nation's energy production and consumption.

Materials
- Lab 4 Data Sheet
- computer with Internet access and spreadsheet software installed

Lab Prep

1. Download the Lab 4 Data Sheet. The first activity on the sheet will help you become familiar with the tools and graphics capabilities of spreadsheet software.

Make Observations

1. Look at the graphic on the data sheet. What is the logic for composing this graphic as a stack of arrows that go from left to right? What do the arrow tails represent? What do the arrowheads represent?

2. Which energy resource accounts for most of our domestic energy production? Which energy resource accounts for most of our imports?

3. Using spreadsheet software, create a chart and associated bar graph that displays the energy supply sources in the United States.

4. Which energy source would be easiest to reduce? Explain.

5. Using spreadsheet software, create a chart and associated bar graph that displays the ways in which the energy supplies in the United States are used.

6. Which of these uses would be easiest to reduce? Explain.

7. What percentage of United States' energy consumption is met through the burning of fossil fuels?

8. What percentage of United States' energy consumption is used for transportation?

9. How effective was displaying your data in a bar graph? Can you present the data with a more effective graphic style? Why would this new style be more effective?

Journal Question How will going green affect a community's energy supply and use?

104 • Opermation: Infinite Potential   www.jason.org

Into the Future

Imagine a future in which energy is abundant and society exists in balance with Earth. What sort of energy strategies and new technologies will safeguard this lifestyle? What will our cities and homes look like?

The Need For Change

Constance Adams understands the need for change, and how a "green" approach to energy can safeguard our planet. This may not mean that we will need to set our home thermostats to uncomfortable temperatures or give up driving cars. What it calls for is changes in the way we act as individuals, countries, and a global society to become more energy responsible. Constance considers this when she designs energy-efficient structures. Depending less upon the burning of fossil fuels, the buildings she designs use solar heating to maintain a comfortable living space. Not only do these structures have less impact on the environment, but her structures depend more upon renewable and inexhaustible resources, such as solar and geothermal energy.

Constance is not alone in creating an architecture for the future. A diverse group of scientists, engineers, and architects are exploring ways in which our society will meet energy demands. They search for affordable generation methods and materials that will have less impact on the planet.

Although the combustion of fossil fuels may one day be replaced by other energy technologies, for now we are very much dependent upon oil and coal. That is why, in addition to developing alternative energy strategies, scientists are exploring better and more affordable emissions-control technologies. These advancements in technology will capture carbon and other pollutants before they enter the atmosphere.

Imagine the Future

Industries are working to increase efficiency in the way we currently generate, use, store, and transmit electrical energy. Not only are these changes better for the environment, but they are becoming more cost competitive.

What will the car of the future look like? How will it be powered? These questions are very much a part of the retooling of the auto industry. Engineers are continuing to improve the efficiency of these vehicles. Did you realize that some vehicles equipped with special fuel-efficient engines can travel from New York to San Francisco on a single tank of gasoline? Although they are not commercially available, these experimental vehicles offer a glimpse into the potential of fuel economy. However, we do not need super efficiency to make a difference. With every increase in efficiency, we extend available supplies and lessen our impact on the environment.

So what will the future—your future—be like? What decisions will you make? Will your behavior change? Will you become involved in shaping energy policy? Perhaps, like Constance Adams, you will enter a career in space architecture. As an architect, you might combine your understanding of energy, structures, and decision making to design a more energy-efficient world.

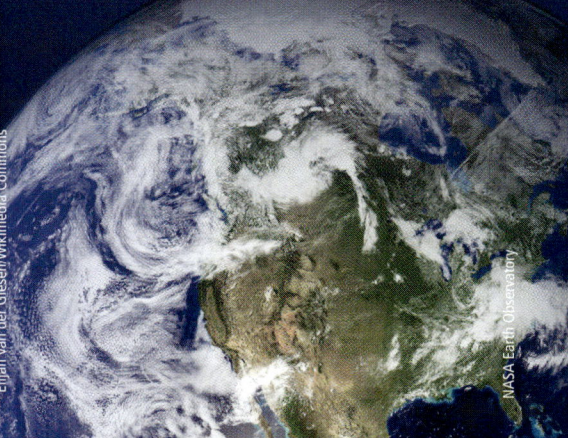

Mission 5: Energy Security—Powering Our Future

Field Assignment

Commencing Countdown

Recall that your mission is to *create a blueprint for a secure energy future*. Now that you have been fully briefed, you will get a chance to see what it feels like to walk in a space architect's shoes. Before Constance even begins a project, she must have a firm understanding of each and every factor that her design must meet. Her challenge was to design a space module that could hold a crew of four astronauts for long-term space travel—flying to Mars and back! There were many factors that Constance and her team needed to think about when designing TransHab.

One important aspect that had to be considered was the materials for her designs. These had to be carefully selected because even the best design will fail if the materials are unsuitable. You will learn, first hand, about the special design considerations and materials selected to make TransHab and other energy-efficient designs and structures Constance has helped build here on Earth. Afterward, you will have the opportunity to apply this knowledge to better equip your home for a future that will be more energy efficient.

Whether your home's energy comes from renewable, non-renewable, or inexhaustible sources, or a combination of the three, you will be challenged to create an action plan that will help reduce the energy consumption of your home.

⚠ **Caution!** Any home modifications need to be approved and supervised by a responsible adult.

> **Objectives:** To complete this mission, accomplish the following objectives:
> - Discover how energy efficiency can be improved through design.
> - Explore a variety of building materials used here on Earth and in space that have energy conservation applications.
> - Create an action plan to make your home more energy efficient.
> - Design an investigation that aims to assess and evaluate your energy-efficient action plan.

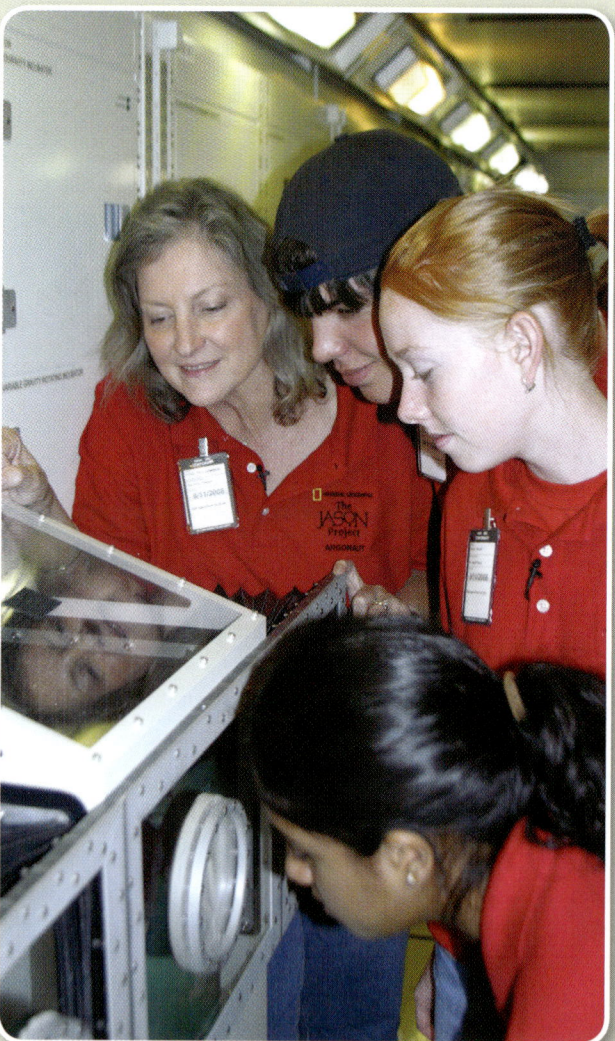

Materials
- Mission 5 Field Assignment Data Sheet
- general household tools and supplies
- variety of architectural building materials provided by instructor
- computer with Internet access

Field Preparation

1. Research a variety of architectural materials designed for energy efficiency.
2. Identify and describe aspects of the materials that allow them to be energy efficient.
3. Explain why these materials could be suitable for space travel.
4. Explain how energy efficiency can be improved through design.

Mission Challenge

1. Create an action plan to make your home more energy efficient by using what you have learned.
2. Have your action plan assessed by a teacher or responsible parent or guardian.
3. Incorporate your action plan. Design an investigation to assess the effectiveness of your action plan. (For example, over several weeks, you could apply your action plan to your home and compare the kW•h used and/or the cost of electricity to last year's energy bill.)

Mission Debrief

Evaluate the effectiveness of your energy efficiency action plan. Compare and discuss your experimental findings with your classmates.

Journal Question TransHab was a collaborative project that took almost a decade to put into orbit. Evaluate the importance of careful planning, testing, and collaboration as an approach to creating new and innovative energy efficient products.

Connections

Weird & Wacky Science

Scientists Take Aim at Creating a Pea-Sized Sun!

Hot off the presses—at over 100 million degrees Celsius: Researchers in California are preparing to create a miniature version of the sun. This tiny, energy-emitting model of our nearest star will be used to learn more about nuclear fusion.

Imagine creating a star within a capsule that is no larger than a pea! Although this sounds like science fiction, it is the future of science for researchers at the National Ignition Facility (NIF). Working at the Department of Energy-funded labs in Livermore, California, these scientists hope to ignite a model of the sun. What they uncover may lead to the accelerated development of nuclear fusion as an affordable and clean energy resource.

In 1952, humankind harnessed the energy of fusion using a destructive device called the H-bomb. Upon detonation, the warhead's content of hydrogen nuclei joined together to form heavier particles. During this process, a small amount of mass was transformed into the massive energy of the explosion.

Over a half century later, scientists have not yet learned to control fusion. We can still detonate a fusion explosion, but we are still unable to harness the energy released to make it useful in electrical energy generation. In contrast, nuclear fission of enriched uranium can be controlled and is currently used in nuclear-fueled power plants.

Perhaps the greatest challenge to harnessing the energy of fusion has been the extreme conditions under which the process occurs. Fusion requires a temperature of over 100 million degrees Celsius (180 million degrees Fahrenheit), hot enough to melt any sort of physical wall that would contain a medium-sized reaction. However, scientists have come up with an interesting approach—confinement using a powerful magnetic field. In addition to its role as a force field, the magnetic energy would also generate a good deal of the thermal energy needed for the reaction. This idea of

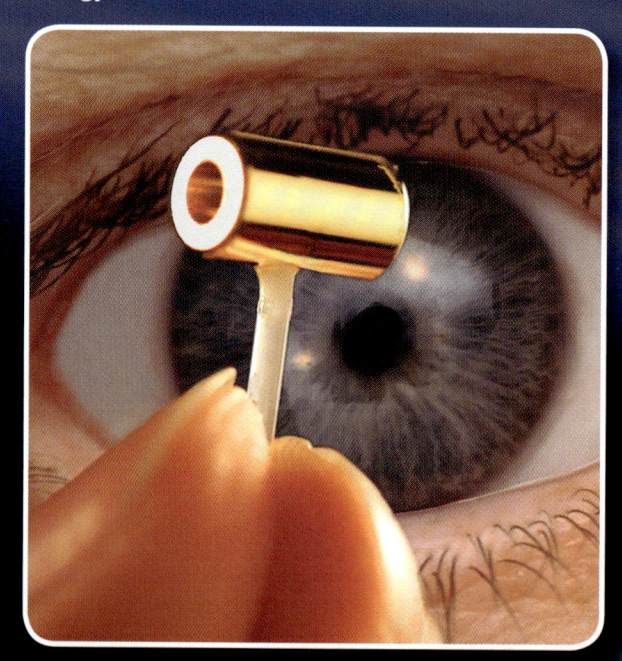

Photos p. 108, p. 109 (top left and bottom): Credit is given to Lawrence Livermore National Security, LLC, Lawrence Livermore National Laboratory, and the Department of Energy under whose auspices this work was performed.

108 • Operation: Infinite Potential   www.jason.org

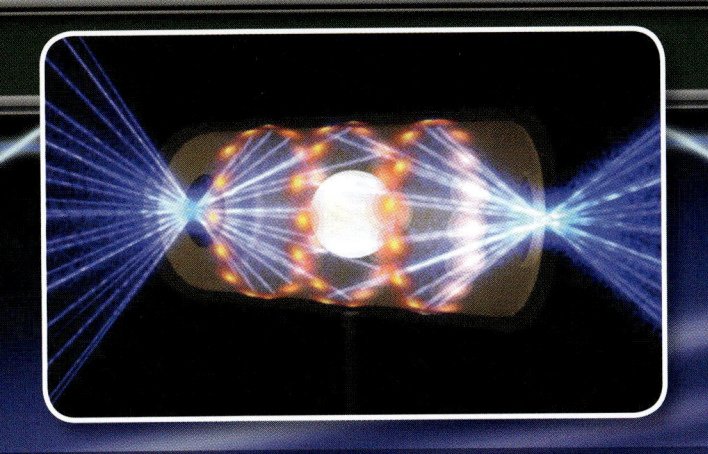

using huge magnetic fields resulted in the development of experimental "tokamak" reactors. Although tokamak designs have been explored for over 50 years, none have successfully produced a sustained fusion reaction.

The scientists at NIF are now exploring a different approach to the fusion process. Instead of using magnetic energy, they are investigating a different energy input—light. To be more specific, they are developing the use of high-powered lasers with the energy of one trillion 100-W light bulbs to trigger this nuclear reaction.

The target of this intense beam of light is a chamber about the size of a small pea. The chamber will be filled with fusible hydrogen. On command, the lasers will blast this chamber with energy causing an implosion. It is hoped that the force of this implosion will be great enough to compress the hydrogen nuclei to the point of fusion. With the right design and ample power, this fusion engine may release as much as 100 times the energy that was put into the system. When sufficient energy is released, the reaction can reach ignition. At ignition, fusion becomes self-sustaining and does not require the continual input of energy for the reaction to continue. Understanding ignition is a major goal of the research.

If we are able to achieve a point where more energy is created through fusion than is required to achieve it, the next step would be using this reaction to generate electrical energy. Engineers plan for a fusion-fueled plant in which a laser would be pulsed several times a second at fuel targets. Attaining ignition, the fuel would generate an enormous amount of thermal energy. Heat would transfer to a surrounding fluid coolant. The heated coolant would then transfer its energy to water, producing steam, which would power a turbine generator. Electricity generated in this process would enter an electric power grid and be distributed as needed.

Courtesy of CRPP-EPFL, Association Suisse-Euratom

Your Turn

Compose a blueprint for an electricity-generating station that is fueled by hydrogen fusion. Label and describe the role of all components involved in the energy transfers and transformations, from the ignition of fusion to home distribution.

Connections: Weird & Wacky Science • 109

The JASON Project Argonaut Program

Join the Argonaut Adventure!

Work with and learn from the greatest explorers, scientists, and researchers in the world as they engage in today's most exciting scientific explorations. JASON is always looking for Argonauts to join our science adventure. Find out here how you can be part of the team!

Local Argonauts

Local Argonauts work with other students in their classrooms and communities to explore and discover the wonders of science.

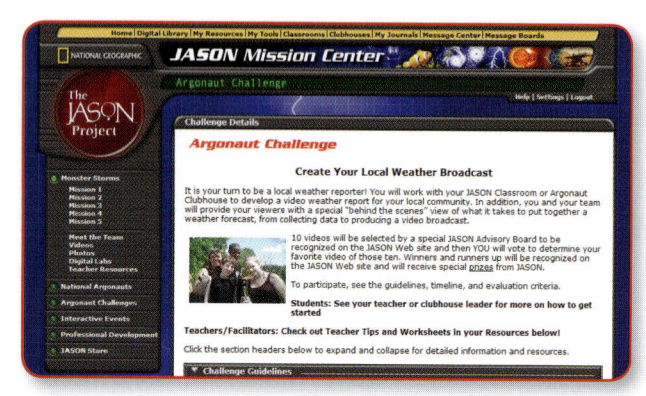

▲ Students will find complete directions and numerous helpful resources for completing the Argonaut Challenge in the **JASON Mission Center**.

Take the Argonaut Challenge

The Argonaut Challenge is an interactive science activity that gives you the opportunity to produce a multimedia project and share it with the entire JASON Community. Are you up for the Challenge? Go to the **JASON Mission Center** and find out about this year's challenge!

Interact with JASON Argonauts

The **JASON Mission Center** is your gateway to meeting the JASON National Argonauts from *Operation: Infinite Potential.* Log into the **JASON Mission Center**, click **Message Boards**, ask the Argos a question, and they will be happy to respond! You will also be able to discuss JASON with other students from around the world.

The **JASON Mission Center** is home to live and interactive events with Argonauts such as Argo podcasts, chats, and other events. Be sure to log on to learn more about when the events are happening and how you can submit your questions to be answered.

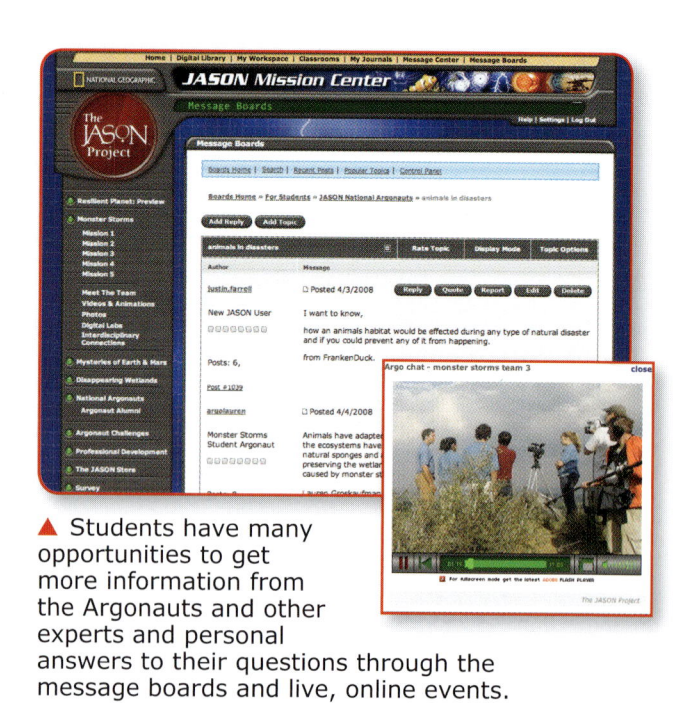

▲ Students have many opportunities to get more information from the Argonauts and other experts and personal answers to their questions through the message boards and live, online events.

Begin your Argonaut Adventure at www.jason.org

National Argonauts

Each year, an elite group of National Argonauts ventures into the field to work with JASON Host Researchers on timely and exciting science explorations. Go online and check out their profiles, journals, photo essays, and videos from the field. See what it is like to be a National JASON Argonaut!

Interested in becoming a National Argonaut yourself and working with the next group of JASON Host Researchers? Check online often to learn about the next opportunity and how to apply!

▲ The Argos experience energy transfers and transformations while white water rafting in West Virginia.

▲ Argonauts work on carbon-control problems by understanding how electricity can be generated.

▲ The Argos work with alpaca scat to understand how biofuels can be created.

National Argonaut Alumni

Many JASON participants embody the idea of life-long learning through a journey of exploration and discovery. These Argonaut Alumni profiles show how JASON has enriched and influenced the lives of students who have gone on to be scientists, explorers, and researchers in their own right. The JASON Project helped them excel not only academically, but also in their lives at home and in their communities. Maybe JASON will make a difference in your life, too.

Read the alumni profiles by clicking **Argonaut Alumni** in the **JASON Mission Center**.

▲ (Left) Kristin Ludwig as a Student Argonaut in 1993 with Dr. Robert Ballard, and (right) studying deep-sea Oceanography at the University of Washington.

The JASON Project Argonaut Program • 111

Meet the Team

Each year JASON recruits a team of expert scientists, students, and teachers to serve on our Missions. The team includes a Host Researcher, Teacher Argonauts, and Student Argonauts like you. To learn more about the Host Researchers, login to the *JASON Mission Center* to read their bios and view the *Meet the Researcher* videos. You can also follow the Argonauts' adventures through their captivating bios, journals, and photographic galleries.

Student Argonauts

HIYAM AÑORVE GARZA
Monterrey, Mexico
Mission 3, Mission 4

Math is Hiyam's favorite subject and she feels high-level math skills can help one solve many of the world's problems.

Hiyam is willing to try any adventure that comes her way, including white water rafting and roller coasters.

HANNAH ZIERDEN
Cardington, OH
Mission 4, Mission 5

Hannah is a leader both in and outside of school. She has held many school offices over the years. She also volunteers with the Big Buddies program.

People who may not have a huge amount of ability, but who attempt things and give it their all inspire her. She thinks that 100 percent of one's effort should be given on every endeavor.

ANTHONY LOPEZ
Charlotte, NC
Mission 5

Anthony has developed leadership skills through classroom experiences and volunteer activities such as Boy Scouts. In Boy Scouts, he has earned the World Conservation Award.

Anthony enjoys playing rugby and spending time in his kitchen conducting culinary experiments.

JOEY BOTROS
Wichita, KS
Mission 1, Mission 4

Joey has a specific interest in cellular biology and human physiology.

Albert Einstein is Joey's hero. Joey believes Einstein would be all over the energy crisis, trying to find innovative ways to solve the situation our world is currently facing.

MADHU RAMANKUTTY
Vienna, VA
Mission 2, Mission 5

Madhu has taken part in four JASON curricula programs since 2002.

She wants to have a hands-on approach to science so that she can use it in her future. She thinks that JASON's curriculum really focuses on connecting science to other subjects.

TIM WEST
Richmond, VA
Mission 2, Mission 3

Tim enjoys exploring technology and sees the potential of technology research and development for any subject.

He also enjoys applying mathematics to help solve scientific problems. He wants to raise global awareness of marine biology and environmental engineering.

LINDSAY HANNAH
Grand Blanc, MI
Mission 1

Lindsay enjoys biology and was Director/Producer of her school's closed circuit TV programming. Lindsay has been an avid dancer since age three.

Throughout her school career, each one of Lindsay's teachers has told her that she can do anything she puts her mind to!

OLUWATOBA (TOBA) FASERU
South Riding, VA
Mission 1, Mission 2

Toba's career goal is to be a scientist and inventor.

He is constantly asking "why" and "what if?" Toba enjoys astronomy, physics, and chemistry, and was in the all-district and all-region chorus.

JACLYN MARTIN
San Jose, CA
Mission 3

Jaclyn's interest in science started early in life when her uncle gave her a bag of sand. Sifting through it, she found fossilized shark teeth and petrified coral.

When not busy with school, Jaclyn is a childcare volunteer and helps at the Silicon Valley Down Syndrome Network.

YOU ARE AN ARGONAUT TOO!

What are your interests? What would you want to tell other people about yourself? What do you like most about your JASON experience and being part of the JASON community?

112 • Operation: Infinite Potential   www.jason.org

Teacher Argonauts

CYNTHIA "CINDY" PARISH
Beaumont, TX
Mission 5

In addition to JASON, Cindy has participated in a geology field camp and in summer paleontology research in Utah and Big Bend, TX. She is particularly proud of how she can develop enthusiasm for science in students with diverse backgrounds.

JASON has taught her to believe in herself and reach beyond the ordinary for solutions to world problems, such as energy research and new technology.

MELISSA HALL
Sadlier, Australia
Mission 3, Mission 4

Melissa is a teacher leader in her school, helping science teachers conduct experiments and use JASON resources. She has coordinated environmental awareness days at her school and loves bushwalking and caving.

Melissa's desire to learn new things comes out in her teaching, and her passion for teaching is apparent the moment that you meet her.

BRYAN IE
Sydney, Australia
Mission 1, Mission 2

Bryan sees teaching science as an "exciting and dynamic craft." His use of JASON in the classroom is fueled by his desire to provide every learner with relevant, real-world activities that meet the needs of ALL learners.

Being a teacher is something that keeps him young, or at least feeling younger. The energy he can get from students' enthusiasm and endless curiosity makes teaching a truly unique and rewarding profession for him.

Host Researchers

JANET GREEN
Space Weather Physicist, NOAA

Janet is a physicist with an interest in the sun and its impact on satellites, communication systems, and power grids. Before working for NOAA, she was a video game tester, assigned to the task of "breaking" the games she worked with to improve their playability.

VASILY TITOV
Director, NOAA Center for Tsunami Research

Vasily's mathematics background and interest in tsunamis come together as he develops computer models that predict the impact these waves might have on coasts and in harbors around the world. He is very concerned with protecting people from the events he studies.

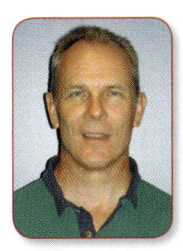

LARRY SHADLE
Research Group Leader, Model Validation Research Group, NETL

Coal has been Larry's professional pursuit. Burning coal more efficiently and cleaning up its emissions has been his focus for his entire career. He was motivated to pursue fuel sciences when President Carter called for the United States to conserve energy and explore alternate forms during an energy crisis.

MARTIN KELLER
Director, BioEnergy Science Center, Oak Ridge National Laboratory

Microbes have been at the center of Martin's research. From the origins of life on Earth to commercial applications in the biotechnology industry, Martin has worked on major research projects in Germany, San Diego and now at ORNL.

CONSTANCE ADAMS
National Geographic Emerging Explorer
Space Architect

From the ISS to potentially Mars, but even back here on Earth, Constance looks to use energy in the most efficient ways possible utilizing local materials, elegant design, and science to make her structures aesthetically pleasing as well as functional.

John Childs/Synthesis International, USA

Find out more about the team by going online: *www.jason.org*

All other photos pp. 112–113 by Peter Haydock, The JASON Project

meet the Team • 113

Math and Building Tools

Build a Convection Detector

Materials
- detector pattern (downloaded from JMC)
- thread
- tape
- scissors
- snap swivel
- vegetable oil
- cotton tipped applicator
- paper towels

Assembly

1. Use your scissors to carefully cut out the spiral design.
2. Obtain one length of thread, about 20 cm long.
3. Use the applicator to brush a small amount of vegetable oil over the rotating shaft of the snap swivel. Work in the oil so that it swivels with as little resistance as possible. Dab off excess oil with paper towels.
4. Tie one end of the thread to the fixed loop at one end of the snap swivel. Note: Do not tie the thread to the clasp.
5. Trim away any excess thread.
6. To test its unrestricted rotation, hold the detector by the clasp end. The spiral should hang freely, pulling the thread taut. Move the detector up and down. Observe the behavior of the spiral.

Math Tools

In their research, scientists use a variety of tools. Some tools, such as microscopes and seismographs, extend the power of observation. Other tools help scientists understand and analyze data. Equations are tools that can be used to reveal unknown values. Below, you will find an assortment of equations that can be used when studying energy in its different forms, transformations, and transfers.

Equations
Mechanical kinetic energy = $\frac{1}{2} mv^2$
Gravitational potential energy = mgh
Work = Fd
Velocity of a wave = $f \lambda$
Ohm's Law: Voltage = IR
Power (electrical) = IV
Energy = mc^2
Power (mechanical) = $\frac{w}{t}$
Average speed = $\frac{\text{total } d}{\text{total } t}$
Efficiency = $\frac{\text{useful energy transferred}}{\text{total energy used}} \times 100$

Measurement	SI Base Unit	Abbreviation
m = mass	kilogram	kg
v = velocity	meter/second	m/s
g = local gravitational acceleration	9.8 meters/second² at sea level	m/s²
h = height	meter	m
F = force	Newton	N
d = distance	meter	m
f = frequency	Hertz	Hz
λ = wavelength	meter	m
V = voltage	Volt	V
I = current	Ampere	A
R = resistance	Ohms	Ω
P = power	Watt	W
E = energy	Joule	J
c = speed of light	3.0x10⁸ meters/second (in a vacuum)	m/s
w = work	Joule	J
t = time	second	s

Build an Electrical Generator

Materials
- small can with cover (can and cover pre-drilled with central hole)
- pencil (without point)
- 4 strong rectangular magnets
- masking tape
- 70 meter spool of 30 gauge magnet wire
- sandpaper

Assembly

1. Wrap the enamel-covered wire around the center of the can, forming a tight coil that is about 5 cm wide. Leave about 20 cm of the wire unwrapped on both ends of the coil.

2. Secure the wrapped coil to the container with strips of masking tape.

3. Use sandpaper to scrape off about 5 cm of enamel from both ends of the wire.

4. Position one magnet along the central axis of the pencil. Place a second magnet on the other side of the pencil, keeping the same orientation and allowing the unlike poles to hold the magnets in place. Add another magnet to each of these inner magnets, allowing the magnetic attraction to maintain each double magnet stack.

5. Use a strip of masking tape to secure both stacks to the central region of the pencil.

6. Pass the pencil and magnet assembly into the wire-wrapped can. Hold the assembly so that the pencil end sticks through the pre-drilled hole. Snap the cover on the can, allowing the other end of the pencil to stick through the lid's opening.

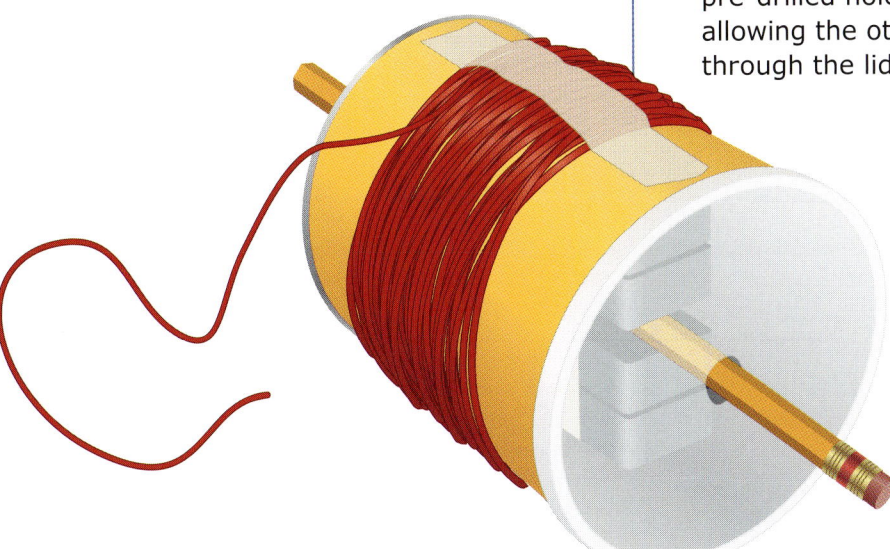

Math and Building Tools • 115

Glossary

A

absorb describes the phenomenon in which electromagnetic energy is retained without reflection or transmission when passing through an object or a substance (16)

active solar heating the collection and use of solar energy with the use of moving mechanical parts (74)

alternating current (AC) an electric current that reverses its direction within a circuit at regular cycles (58)

amplitude the amount of displacement of a wave from the rest position (34)

B

biofuel a renewable carbon-based fuel, such as ethanol from corn (78)

C

carbon footprint a measure of human impact based upon the generation of carbon-based greenhouse gases (66,73)

circuit the complete path of an electric current (56)

circuit breaker a safety switch designed to automatically shut off when a circuit becomes overloaded (61)

coal a type of fossil fuel, coal is a black or brownish black rock formed from the remains of plant life (50,64)

combustion a reaction between a fuel and oxygen, releasing thermal and electromagnetic energy (51)

compact fluorescent lamp (CFL) an energy saving alternative to incandescent lamps (43,97)

compression high density region of a mechanical compression wave where the medium is being squashed (34)

compression wave type of mechanical wave in which matter in the medium moves back and forth in the same direction that the wave travels (33)

concave lens a lens which has a depression in its center (16)

conduction the transfer of thermal energy between atoms and molecules that occurs within an object or between objects that touch (38)

conductor any material which easily transfers energy (40)

Conservation of Energy energy is neither created nor destroyed; it only changes form or is transferred (31)

Conservation of Mass in any chemical reaction, the total mass of the reactants equals the total mass of the products (10)

convection the transfer of thermal energy that occurs by the flow of material (38)

convex lens a lens which has a bulge in its center (16)

coronal mass ejection (CME) a magnetic field that carries particles from the sun, creating a shockwave that accelerates solar wind particles toward Earth (8)

crest high point of a transverse wave (34)

D

direct current (DC) an electric current that flows in only one direction through a circuit (56)

E

efficiency a measurement of how effectively the total energy of a system is transferred and transformed into useful energy (42)

electric current a flow of charges (56)

electrical conductor a material that allows charges (i.e., electrical current) to flow easily (56)

electrical insulator a material, such as wood, glass or plastic, that does not allow charges to flow easily (56)

electromagnet a device consisting of an iron or steel core that is magnetized by electric current in a coil that surrounds it (55)

electromagnetic wave travels at or near the speed of light in a vacuum and does not require a medium through which to travel (12,33)

energy the ability to do work (8)

energy portfolio the energy resource choices available to a location determined by economic, environmental, and technological factors (50)

energy security access to a reliable source of energy (94)

energy transfer a process by which energy remains in the same form but is passed between two or more components of a system (30)

energy transformation a process by which energy changes form within or between components of a system (31)

ethanol type of alcohol created by the fermentation of sugars by yeast (78)

F

force a push or a pull (8,14)

fossil fuel carbon-based fuel, such as coal, oil, or natural gas, formed from plant or animal remains (51,64)

frequency number of waves that pass a given point within a given period of time (34)

fuel a material consumed to produce energy (51)

fuel cell a device which produces electrical and thermal energy from a fuel and an oxidant (85)

fuse a safety device which consists of a thin piece of wire that melts when too much current is sent through it, thereby breaking the circuit and stopping the current from flowing (61)

G

generator a device that transforms mechanical energy into electrical energy (54,58)

geothermal energy energy derived from the heat within Earth (85)

global warming an increase of global temperatures as a result of increased emissions of greenhouse gases, such as CO_2, into the atmosphere (65)

gravity force of attraction that exists between any two objects (9)

ground fault circuit interrupter (GFCI) a safety device that continually detects how much electric current is flowing through a circuit and shuts off power whenever it detects any "leak" in the circuit that could cause an electrocution (61)

H

heat the movement of thermal energy between substances (38)

hybrid vehicle vehicle powered directly or indirectly by two or more energy sources (96)

hydrocarbon an organic compound that consists only of hydrogen and carbon atoms (64)

hydroelectric energy electrical energy that is generated as a result of water spinning a turbine (62)

I

inexhaustable describes an energy source, such as the sun or geothermal, which has an almost unlimited supply (72)

infrared (IR) electromagnetic energy with wavelengths slightly longer than visible light but shorter than microwaves (18)

insulator any material which poorly transfers energy (40)

internal combustion engine type of engine in which a fuel, such as gasoline, is combusted inside of the combustion chamber of an engine resulting in the movement of a piston (53)

––––––––––––––––––––– J –––––––––––––––––––––

joule SI unit of energy or work. 4.18 joules is equivalent to one calorie. (14)

––––––––––––––––––––– K –––––––––––––––––––––

kinetic energy (KE) energy of motion (8)

––––––––––––––––––––– M –––––––––––––––––––––

magnetic field the region surrounding a magnet where magnetic forces are exerted (8,54)

mechanical wave any wave which requires a medium to transfer energy (33)

––––––––––––––––––––– N –––––––––––––––––––––

natural gas a colorless and odorless gas formed from the remains of animals and plants (64)

non-renewable a fuel source, such as coal, oil, or natural gas, which once used, cannot be easily reused or recreated (61,73)

nuclear fission the process of splitting atoms of a radioactive material, such as uranium, in order to release energy (61)

nuclear fusion a process in which two or more atomic nuclei are joined together to form a single nucleus with a slightly smaller mass than the sum of their original masses (83)

––––––––––––––––––––– O –––––––––––––––––––––

Ohm's Law Ohm's Law states that the current through a conductor between two points is directly proportional to the voltage across two points, and inversely proportional to the resistance between them (56)

oil a yellow to black liquid formed from the remains of plants and animals that died in a marine environment and that were subsequently subjected to extreme heat and pressure for millions of years (64)

opaque property of an object which light does not pass through, but is reflected off its surface (16)

––––––––––––––––––––– P –––––––––––––––––––––

passive solar heating the collection and use of solar energy without the use of moving mechanical parts (74)

petroleum a classification of liquid hydrocarbon mixtures—includes crude oil and natural gas (52)

photovoltaic (PV) technology focused on transforming electromagnetic energy into electrical energy (76)

photovoltaic array multiple photovoltaic cells linked together to create higher energy output (76)

potential energy (PE) stored energy not yet in motion (8)

power the rate at which work is performed (14)

power distribution grid the system of facilities and structures that provides an organized and efficient way to deliver electrical energy from the power plant to all of the people and things that need it (60)

––––––––––––––––––––– Q –––––––––––––––––––––

quad unit used to measure large amounts of energy, equal to 1,000,000,000,000,000 (10 to the 15th power) Btu (93)

––––––––––––––––––––– R –––––––––––––––––––––

radiation the transfer of energy that occurs by the propagation of electromagnetic waves (38)

rarefaction low density region of a mechanical compression wave where the medium is being stretched (34)

ray diagram drawing that illustrates the behavior of light (16)

reflect occurs when a wave strikes the boundary between two mediums and bounces off (16)

refract bending of a wave as it moves from one medium into another medium (16)

renewable an energy source, such as water or biofuel, which even after being used, can be easily regenerated (62,72)

resistance the measure of how easily charges flow through a substance or device (56)

––––––––––––––––––––– S –––––––––––––––––––––

smart electric grid digital distribution and management system that connects sources of energy generation to a variety of end users using superconducting transmission lines (100)

solar wind continual torrent of particles and magnetic field emitted by the sun (22)

system set of components and processes related by energy transfers and transformations (30)

––––––––––––––––––––– T –––––––––––––––––––––

temperature the measurement of the average kinetic energy of the particles within a sample of matter (12,38)

thermal energy the total energy content of a system (12,38)

tide the continuous rise and fall of Earth's ocean surface caused by gravitational forces between the sun, moon, and Earth (82)

TransHab Module an inflatable space structure that expands to create additional living space (92)

translucent property of an object allowing light to pass through, but scattering the ray pattern (16)

transmit describes the phenomenon in which electromagnetic energy is able to pass through an object or a substance (16)

transparent property of an object allowing light to pass through with little change to the pattern of rays (16)

transverse wave type of mechanical wave in which the wave energy causes matter in the medium to move up and down or back and forth at right angles to the direction the wave travels (34)

trough low point of a transverse wave (34)

turbine a device that consists of a series of blades arranged in a circle, which may be spun around by the introduction of a fluid such as steam pushing against the blades (53)

––––––––––––––––––––– U –––––––––––––––––––––

ultraviolet (UV) electromagnetic energy with wavelengths slightly shorter than visible light but longer than x-rays (18)

––––––––––––––––––––– V –––––––––––––––––––––

voltage the difference in electric potential between two locations (56)

––––––––––––––––––––– W –––––––––––––––––––––

wave a progressive disturbance that transfers energy from one location to another (12,33)

wavelength for a transverse wave, distance as measured by the tops of two adjacent crests or the bottoms of two adjacent troughs; for a compression wave, distance as measured by the centers of adjacent compressions or rarefactions (34)

wind farm a group of wind turbines that work together to produce higher levels of electricity (80)

wind turbine a structure used to transform wind energy into electrical energy using a rotor and a generator (80)

work energy used to move an object a certain distance using a force (8,14)

Glossary • 117

Credits

The JASON Project would like to acknowledge the many people who have made valuable contributions in the development of the *Operation: Infinite Potential* curriculum.

Partners
National Geographic Society
National Aeronautics and Space Administration (NASA)
National Oceanic and Atmospheric Administration (NOAA)
United States Department of Energy (DOE)
National Energy Technology Laboratory (NETL)
Oak Ridge National Laboratory (ORNL)

Supporting Sponsor
Shell

Host Researchers
Constance Adams, Space Architect and National Geographic Emerging Explorer, Houston, TX
Dr. Janet Green, Space Weather Physicist at Space Weather Prediction Center NOAA, Boulder, CO
Dr. Martin Keller, Division Director, BioEnergy Science Center, Oak Ridge National Laboratory DOE, Oak Ridge, TN
Dr. Larry Shadle, Research Group Leader at National Energy Technology Laboratory DOE, Morgantown, WV
Dr. Vasily Titov, Director of NOAA Center for Tsunami Research, Pacific Marine Environmental Laboratory, Seattle, WA

Teacher Argonauts
Melissa Hall, Sadleir, Australia
Bryan Ie, Sydney, Australia
Cynthia "Cindy" Parish, Beaumont, TX

Student Argonauts
Hiyam Añorve Garza, Monterrey, Mexico
Joey Botros, Wichita, KS
Toba Faseru, South Riding, VA
Lindsay Hannah, Grand Blanc, MI
Anthony Lopez, Charlotte, NC
Jackie Martin, San Jose, CA
Madhu Ramankutty, Vienna, VA
Tim West, Richmond, VA
Hannah Zierden, Cardington, OH

JASON Volunteer Corps
Kim Castagna, Executive Committee Chair
Mary Cahill, Executive Committee Co-Chair
Krystyna Plut, Participation and Review Committee Chair
Dee McLellan, Participation and Review Committee Co-Chair
Steve Jarman, Recognition Committee Chair
Marjorie Sparks, Recognition Committee Co-Chair
Marti Dekker, Communication and Membership Committee Chair
Karen Bejin, Communication and Membership Committee Co-Chair

Special Thanks
Don Caminati, NASA – Johnson Space Center
Allen McDonald, Fort Martin Power Station
Mike Nowak, National Energy Technology Laboratory
Mr. Todd T. Worstell, National Energy Technology Laboratory
Mr. Keith B. Knotts, National Energy Technology Laboratory
Dr. Chris Guenther, National Energy Technology Laboratory
Mr. Larry A. Kincell, National Energy Technology Laboratory
Mr. Michael J. Monahan, National Energy Technology Laboratory
Ms. R. Diane Newlon, National Energy Technology Laboratory
Suzy Tichenor, Oak Ridge National Laboratory
Dr. Thomas Zacharia, Oak Ridge National Laboratory
NASA – Johnson Space Center
NOAA Office of Education
NOAA – Space Weather Prediction Center
Nuclear Energy Institute
National Energy Technology Laboratory
Oak Ridge National Laboratory
Smithsonian Air and Space Museum
University of Colorado – Laboratory for Atmospheric and Space Physics
U.S. Rep. Alan B. Mollohan
Sandi Brallier, Marketing Assistant
Michael DiSpezio, Consultant and Writer
Christopher Houston, Flash Developer
Penny Kline, Business Coordinator
David Shaub, IT Systems Administrator
Katie Short, Technology Project Manager
John Stafford, Lead Developer

The JASON Project
Caleb M. Schutz, President
Michael Apfeldorf, Director, Professional Development
Grace Bosco, Executive Assistant
Tammy Bruley, Business Manager
Whitney Caldwell, Manager, Professional Development, Argonaut Program
UT Chanikornpradit, Senior Applications Developer
Lee Charlton, Logistics Coordinator
Marjee Chmiel, Director, Digital Media
Lisa Campbell Friedman, Director, Media Production
John Gersuk, Executive Vice President, Government Relations
Peter Haydock, Vice President, Curriculum and Professional Development
Bryan Ie, Content Producer, Curriculum
Bill Jewell, Vice President, Digital Media and Technology
Ryan Kincade, Senior Applications Developer
Laura Lott, Chief Operating Officer
Josh Morin, Senior IT Systems Administrator
Arun Murugesan, Applications Developer
Andre Radloff, Content Producer, Curriculum
Orion Smith, Online Content and Community Specialist
Sean Smith, Director, Systems and Projects
Lisa M. Thayne, Content Producer, Curriculum

The JASON Project Board of Trustees
Robert D. Ballard, Ph.D., Founder and Chairman, The JASON Project
John M. Fahey, Jr., President and CEO, National Geographic Society
Terry D. Garcia, Executive Vice President for Mission Programs, National Geographic Society
Christopher A. Liedel, Executive Vice President and Chief Financial Officer, National Geographic Society
Caleb M. Schutz, President, The JASON Project

Operation: Infinite Potential Reviewers

Laura Amatulli, Rochester Hills, MI • **Karen Bejin,** Chippewa Falls, WI • **Ginny Brackett,** Winslow, ME • **Edward J. Brewer,** Berea, OH • **Mary C. Cahill,** McLean, VA • **Kim Castagna,** Ventura, CA • **Kathy Coffey,** Floral Park, NY • **Marcie Colahan,** Burns, OR • **Donna Costa,** New Castle, DE • **Victoria B. Costa,** Riverside, CA • **Marti Dekker,** Zeeland, MI • **Chris Donovan,** Tucson, AZ • **Sara Dykstra,** Navarre, FL • **Tom Fitzpatrick,** Roanoke, VA • **Doug Glasenapp,** Milwaukee, WI • **Dr. Lisa Marie Gonzales,** San Jose, CA • **Jackie Hackman,** Warrenton, VA • **Victoria M. Kehoe,** Aurora, IL • **Laurie Laubacher,** Amherst, OH • **Mellie Lewis,** Key Largo, FL • **Susan Mills,** Peachtree City, GA • **Linda O'Brien,** Sam Rayburn, TX • **Mary Obringer,** New Washington, OH • **Jone Kimberly Preston,** Vero Beach, FL • **Ann Robichaux,** Houma, LA • **Tanya Shank,** Charlotte, NC • **Dawn Sherwood,** Highland Springs, VA • **Carla-Rae Smith,** Colorado Springs, CO • **Susan Stock,** Webster City, IA • **Amy Strong,** Hutchinson, KS • **Chris Taylor,** Boise, ID • **Alaine Tingey,** Las Vegas, NV • **Teresa Tollison,** Laurens, SC • **Vickie Weiss,** Grand Blanc, MI • **Jeni Kocher Zerphy,** Annapolis, MD